THE APOSTLE PROJECT

May Your Adventure Be Blessed

ISA. 40:31

THE APOSTLE PROJECT

Discovery of Matthew and the Frankincense King

DR. JIM RANKIN

XULON PRESS

Xulon Press
2301 Lucien Way #415
Maitland, FL 32751
407.339.4217
www.xulonpress.com

Edited by Ken Dorhout

Printed in the United States of America.

ISBN-13: 9781545670729

Dedicated to the many whom have stood beside me, gone
before me, and had the guts to go where others would not.
To my wife and children for the love to pursue forward.
And most importantly to my Lord and Savior, Jesus Christ;
thank you for your everlasting love and direction.
This is just the beginning of the beginning
of the adventures to come.

Contents

ACKNOWLEDGMENT

WHEN YOU STEP OUT into a journey of faith there are many who come along in that journey, while many will stay behind. The expedition and discoveries you are about to embark upon within these pages are miraculous and there were many who have given up the tradition of the world to see their paths opened up to take part in this journey.

My wife Sherri has been the one who has stood back and watched this develop. She has stepped in when I was frustrated and encouraged me to move further. And finally, she was the spark that pushed me to finish this work, especially after she was able to experience the place and the feelings of discernment while watching all of the evidence come together. My boys and our daughter-in-law (Skyler, Austin and Tensley) have always been there to support me for the cause, and the mission that God has sent me on.

The team in Ethiopia has been a stable force of research and strength through this process. Sisay Tsegay stood beside me when I was doubtful we would never be able to piece this together, while our young friend Asgedom Tazeze and his excitement along the journey was an inspiration. Zemichael Berhane in Axum has always been a friend and a direct link to my friendship with the Guardian. Bante Mlhvet in Gondar has been there through many adventures, and will be there for many more. My friend Aba Like-Likawint-Ezera-Hadis-Yemetsahft-Gubena-Memhir, a high-ranking priest in Gondar, and Asnake Tadesse from Lalibela have been key parts to sharing a great deal of time finding answers to my questions. And then Misgana Genanew and his continued friendship has been enough to encourage me when all seemed to be a dead end.

Those standing alongside our work and our research in the United States has been a true blessing as well. Bev Regehr has been a mainstay in the work both in the U.S. and in Ethiopia, and is there for a constant encouragement. Tim Moore has seen this progress over the years and has that desire to see the truth presented as he has seen first-hand the evidence that will open up to you within these pages. Kathryn Pearce is another who has prayed for our work and traveled to the other-side if the world to be a part of it all. And to Ken Dorhout for his desire to help us with editing and suggestions to help us along the way, even when my mind was at a standstill. I can't go without acknowledging a man who has been there to touch my life without even knowing it, Dr. Verlis Collins. He has always been a true mentor to me and always has words of wisdom from the Scriptures to get me to the next step in my journey.

Then there is a list of many others who God has put in my path to be able to either encourage me along the journey, research the work, or just to say a prayer for the path we've traveled upon. To all of you, thank you for trusting in the leading of God, knowing the final result would be miraculous.

Chapter 1

Matthew: A Journey of Biblical Adventures

ANCIENT ROADS LEAD TO significant events...and those events led us to the truth. The story of Matthew has always been an interesting twist to the men Jesus chose as His disciples and for those selected to complete the writings of the Bible.

Matthew was clearly an unlikely choice. When we really step back in our lives, aren't we just as unlikely?

As I began this journey, I never dreamed it would take such twists and turns. But, yet again, those twists and turns have brought us to an unbelievable truth that only God could have orchestrated. A journey that had to be much like the disciples themselves, turning from the tasks they had been dealt in life, and turning to the mission that Jesus laid out before them.

Think about this, a tax collector turned missionary for the Lord. Wait a minute, that's what happened to me! Well, not a tax collector, but the entertainment field, to pastor, to missionary isn't much different, really. Let's take a look at how this journey began. For starters, I can remember a vivid conversation after a message that a former pastor of mine had preached one Sunday morning. It was dealing with the sufferings one must go through to serve God. He used examples of those who were martyred for their faith in the Bible. During our conversation, I can remember my discussion with the pastor regarding where the information for those martyred can be found. His response was simply, "You can research, and you'll find it."

Flash-forward five years; I'm in Bible seminary, and one of the books we were required to read was *Foxe's Book of Martyrs (ref. 2)*. This book was known as the *Actes and Monuments*, and was published first in 1563, written by John Foxe *(ref. 3)*. When Foxe began his work, it was in the simple form describing the history of martyrdom in the church of the time due to the hands of the Romans, primarily in England and Scotland. As the pages and writings expanded, this book continued on with those of biblical text and their suffering that had been recorded and kept through time, as accumulated by Foxe's research into recorded manuscripts over the years. Much of that information and accounts are gone, but thankfully Foxe (and a few others) kept these texts recorded for our knowledge later *(ref. 4)*.

Other writings prior to Foxe's book were used as well in order to record what had happened historically during the times of these persecutions and killings of the Christians in the early works of the church. As time continued on, writings were lost, destroyed, or simply withered away. Before they disappeared completely, many of the accounts were collected and kept by men who knew this would be helpful to future generations, while others of the modern age dispute the accuracy of these writings. We are about to dive into these writings, which will clearly show that the men of old were closer to the accuracy of what the events truly were, and thankfully recorded them for us to see.

One of the early recorders of the days following the apostles included Clement of Rome (35-99 A.D.), who actually worked alongside the Bible's Peter and Paul, as mentioned by Paul in Philippians.

> *And I intreat thee also, true yokefellow, help those women which laboured with me in the gospel, with Clement also, and with other my fellowlabourers, whose names are in the book of life. Philippians 4:3*

Clement was one of the early church bishops in Rome, and his writings, the Epistles of Clement, included information of the whereabouts of some of the works and fates of the apostles and evangelists *(ref 5)*.

Others who were recording additional history around the time of the apostles included Papias of Hierapolis (60-163 A.D.), who lived during the days of the original apostles and was a disciple of John. Papias, a Greek bishop and author, was one of the earliest known scholars to actually put the oral presentations of the canonical Gospels on paper in his now partially lost writings of *Expositions of the Sayings of the Lord*. There were many others, including Clement of Alexandria (153-217 A.D.), Iranaeus the bishop of Lyon (120-202 A.D.), Origen (185-254 A.D.), and, of course, Jerome (342-420 A.D.) who was one of the most well-known biblical translators and scholars.

One of the earliest to ever collect the happenings of the apostles was Eusebius (265-340 A.D.). In 325 A.D., he wrote the best collection of the complete fates of the men who went out to share the gospel in the early days. Many claimed that Eusebius collected these stories from other church leaders over time. This may be true, and thankfully so, considering that most of the Scriptures we have today were passed down orally at first and later recorded in written form. These writings added to the biblical texts, as many of the ancient manuscripts that we will discuss later, gave finality to the Scriptures of what the Bible only touches upon. For example, Eusebius adds to the story of James' martyrdom from Acts 12:2 with his manuscript of, *The Ecclesiastical History of Eusebius Pamphilus (ref. 6)*, that when James was led to his martyrdom, the man who was leading him was deeply moved as he saw the apostle's witness and confessions that he was a Christian. This was quoted to him by Clement of Alexandria; it was common to pass stories from scholar to scholar and from age to age. There are others involved in recording these events, and we will mention them and their works later in these writings.

Many pastors have referenced the fate of the apostles for years in churches across the globe, but in reality, most don't know where any of this truly comes from. That's what I hope to open up within these pages. When it's simply referenced in a Sunday morning message, most in a congregation just assume it's in the Bible. Since the majority of church-goers really never read the Bible fully, the assumption just seems to stick. Thus, the pastor never has to explain where he heard that information from. And believe me, I'm just as guilty of this from my time in the pulpit. It's something that was passed to you and then you share it from there. It wasn't until later that my research grew. We pray you'll see a clear picture of this throughout this writing before we're finished.

Looking back, I can see that God truly was lining this up for me well before my knowledge was even being tested. Following my first round of Bible college, I truly fell in love with biblical history and the truth of where certain events actually came from. This continued on with me during my years as a pastor, and thankfully, the years following working out in the mission field. Without this previous knowledge, what we discovered would have just been brushed under the rug and forgotten because I would have not understood what was being laid in my path.

THE BIBLE'S MATTHEW

So, who is this man who wrote the God-inspired book? Matthew is much the same as many of the other disciples that Jesus chose. He would not have been a likely choice in the alignment of who we would think would have been more prudent picks. Think about this: Matthew worked for the Romans, the same evil empire that judged and would later crucify the One that he would be following. I strongly believe that Matthew is no different than any of us. God chose His disciples from the most unlikely to take up the cross and follow Him. That's what He does to each of us only if we are willing to fully leave the life we have become accustomed to and turn to a life of serving only Christ.

The Bible gives us a beautiful look into this man named Matthew. We see many aspects of him in the Scriptures, but we also see the truths in the beginning of his work with Jesus. So, let's go to the Bible to get a closer look at this man who gave up all to follow the Savior.

When we first see Matthew within his writing, we learn he was from Capernaum:

> *"And when Jesus was entered into Capernaum,"*
> (Matthew 8:5a)

Jesus' first encounter with Matthew happened in the most unlikely of ways.

> *"And as Jesus passed forth from thence, he saw a man, named Matthew, sitting at the receipt of custom: and he saith unto him, Follow me. And he arose, and followed him." (*Matthew 9:9)

> *"And after these things he went forth, and saw a publican, named Levi, sitting at the receipt of custom: and he said unto him, Follow me. And he left all, rose up, and followed him."* (Luke 5:27, 28a)

Matthew was sitting and collecting taxes of the people; he was publican, a tax collector. This man would have been despised by the local people because of the tactics used to collect the taxes. Many times, the people were over-taxed, and the collectors filled their pockets with the overage. He worked directly for the Roman government, and Matthew would have been a somewhat wealthy man.

When it comes to unlikely, Matthew was a clear choice to fit that description. At the time Jesus had come to him, He had just completed a series of miracles and was making His way through the streets of Capernaum. He saw a Jew named Matthew (Levi), and stopped immediately to approach him. Without hesitation, Jesus moved right to the point. As He did many times, Jesus simply said to them, "*Follow me*" which in

Matthew's case, as with the others, he simply dropped everything and followed the Messiah. There had to be a feeling, an aura, around Jesus. There had to be a supernatural effect on these men to simply stand-up, pack-up, and risk all to follow a man who simply walked up and said, "*Follow me.*"

WHAT'S IN THE NAME?

As we've seen before in Luke 5:27, Matthew is also called Levi, as stated in the Gospel of Mark:

> "*And as he passed by, he saw Levi the son of Alphaeus sitting at the receipt of custom, and said unto him, Follow me. And he arose and followed him.*" (Mark 2:14)

In both Scriptures, Matthew is referred to as Levi by the other Gospel writers, most likely out of respect possibly as his Greek name, but in Matthew's Gospel he shares his name as we see it within the title.

Mattityahu is Matthew in Hebrew meaning, "gift of YAHWEH," or the "gift of the Lord." Many believe this name may have set Matthew apart as Jesus approached him, but regardless, this was the right man for what was to come. Jesus knew His team of missionaries and evangelists well before He began His journey. He knew the importance, the endurance, and desire of each man. It this case, Matthew (Levi), was a gift, as we all are, to serve and share the Word and the path to follow toward Jesus.

For Matthew, as the others had done, he simply walked away from a life of privilege and favor. But one can't help but think that maybe Jesus knew of the pressures upon this man as well. Unlike many of the others chosen to follow the Messiah, Matthew was financially set, and as they say, *with great power comes great responsibility*. As a tax collector, Matthew had to have been under extreme pressure by the Romans to keep the necessary dollars coming in, while at the same time, keeping up his lifestyle. Jesus saw something in this man. He knew this was the man He wanted, and this is the man that would follow and then go out into the world to complete the journey that had now been set before him.

THE REPUTATION IS TESTED AT THE PARTY

As we've seen, Jesus asked Matthew to leave the world he knows, to pack up, and simply leave it all behind to learn, follow, and serve under The Master. Matthew was thrilled with his new calling. How do we know that? It's simple - he threw an extravagant party for Jesus at his home.

> *"And it came to pass, as Jesus sat at meat in the house, behold, many publicans and sinners came and sat down with him and his disciples."* (Matthew 9:10)

> *"And it came to pass, that, as Jesus sat at meat in his house, many publicans and sinners sat also*

> *together with Jesus and his disciples: for there were many, and they followed him." (*Mark 2:15)

> *"And Levi made him a great feast in his own house: and there was a great company of publicans and of others that sat down with them."* (Luke 5:29)

There is no doubt Matthew was among the elite in Capernaum, as we see when he decided to throw a party for his new calling and for Jesus. Many of those who attended were fellow tax collectors, politicians, and others considered to be in sin for their life and actions. In this time period, the Pharisees, the so-called religious hierarchy, came to the party, not to feast, but to criticize Jesus for his accompaniment with such a sinful group of people.

> *"But their scribes and Pharisees murmured against his disciples, saying, Why do ye eat and drink with publicans and sinners?"* (Luke 5:30)

> *"And when the scribes and Pharisees saw him eat with publicans and sinners, they said unto his disciples, How is it that he eateth and drinketh with publicans and sinners?"* (Mark 2:16)

> *"And when the Pharisees saw it, they said unto his disciples, Why eateth your Master with publicans and sinners?"* (Matthew 9:11)

When we look at this interaction, we can quickly notice the cowardness of the Pharisees, as they did not approach Jesus, but His disciples in all three accounts. They condemned Him for gathering, socializing, and feasting with this massive group of sinners. But because Jesus knew why they were there, He simply turned the accusations back onto the Pharisees.

> *"And Jesus answering said unto them, They that are whole need not a physician; but they that are sick. I came not to call the righteous, but sinners to repentance."* (Luke 5:31, 32)

> *"When Jesus heard it, he saith unto them, They that are whole have no need of the physician, but they that are sick: I came not to call the righteous, but sinners to repentance."* (Mark 2:17)

> *"But when Jesus heard that, he said unto them, They that be whole need not a physician, but they that are sick. But go ye and learn what that meaneth, I will have mercy, and not sacrifice: for I am not come to call the righteous, but sinners to repentance."* (Matthew 9:12, 13)

Not only did Jesus enlighten the Pharisees with the God-driven response of coming to lead sinners to repentance, but He opened the hearts and the eyes of those men that had gathered to hear Him:

> *"for there were many, and they followed him."* (Mark 2:15)

Only Mark shares in his Gospel that all of those that came and gathered at Matthew's house gave repentance, and they followed Jesus from that point on. This had to be an eye-opening day for this particular community. Those of wealth, those whom they despised, those whom have cheated the people, all gave up their lives to follow the message of the Good News of Jesus.

WRITER CONFIRMED BY THOSE OF THE AGE

Unfortunately, over time man has begun to believe that they know more about all things, even better than those who were alive when the events actually occurred. With Matthew, this is no exception. The Bible gives us a glimpse into believing those of the age rather than those who came along later. An example of this is in Romans:

> *"For whatsoever things were written aforetime were written for our learning, that we through patience and comfort of the scriptures might have hope."* (Romans 15:4)

And then again in Deuteronomy.

> *Remember the days of old, consider the years of many generations: ask thy father, and he will*

> *shew thee; thy elders, and they will tell thee."* (Deuteronomy 32:7)

Unfortunately, the battles continue in our modern age when man's arrogance presses forward to disprove what those of the past have already confirmed. The Bible tells us to look to those of the past for the truth of the past, but we continue to believe we know more. This has led to more and more confusion to those seeking the Word of God. So, with that said, who is the author of confusion? Obviously, the answer is Satan himself. My purpose of saying all of this is to encourage you to believe what has been written and confirmed when it aligns to God's Word and not someone of the modern age, who simply believes they know best with a new idea - it's all the agent of confusion.

So, who wrote the Gospel of Matthew? Those who lived nearer to the times of the actual events have already confirmed that Matthew was the writer, and this confirmation dates as far back as Papias of Hierapolis, whom we spoke of earlier. Papias' writings, in approximately 100 A.D., were some of the earliest writings of confirmations of the Scriptures, including his verification that Matthew was the writer of Gospel. This was also confirmed through the writings of historian Eusebius of Caesarea in 320 A.D. as he quoted Papias of Hierapolis recognizing Matthew as the writer of the Gospel *(ref. 7, 8, 9, 10)*.

Origen of Alexandria was a theologian and a very early Christian scholar, and he also made reference to Matthew as the author who wrote the gospel *(ref. 11)*. During several of these early writings confirming Matthew, especially in Papias' manuscripts, it was claimed that the Gospel of the apostle was written in Hebrew in Jerusalem and later translated into the Greek text *(ref. 12)*. These men of the days of old also state that the originals were lost in time, but the Gospel we have today was taken from the oldest known manuscripts and are accurate.

One final thought; when Jerome began to collect manuscripts to translate the early Gospels, it was recorded that he received a transcribed copy of Matthew's Gospel directly from the Nazarenes *(ref. 13,14, 15, 16, 17, 18)*. This is another confirmation that the men in the days of the ancients knew the Word of God and could confirm the writers, the text, and had more of an inside advantage than any other modern-day scholars could even dream of having.

Chapter 2

The Foretelling of the Future

THE BEGINNING

LET'S SET THE STAGE. Before we go any further, we must go back to a time earlier than Matthew's addition to Jesus' twelve. First, we go back to the time of the birth of Jesus in Bethlehem. Why begin here? It's a good question, but to truthfully answer this, we have to wait until later. Trust me, it's worth the wait!

It had been a long trek through rough terrain, over mountains, and along dusty caravan roads for Mary, Joseph, and probably many others, in order to reach Bethlehem. The reason for the journey, as we see in the Scriptures, is that a decree from Caesar Augustus stated that all should return to the city of their lineage to be counted for taxation.

> *And it came to pass in those days, that there went out a decree from Caesar Augustus, that all the world should be taxed. (And this taxing was first made when Cyrenius was governor of Syria.) And all went to be taxed, every one into his own city. And Joseph also went up from Galilee, out of the city of Nazareth, into Judaea, unto the city of David, which is called Bethlehem; (because he was of the house and lineage of David:) To be taxed with Mary his espoused wife, being great with child.* (Luke 2:1-5)

This journey was made even more difficult because Mary, as the Scriptures also state, was bearing a child that was supernaturally placed within her by God. As they arrived in Bethlehem, we know the family of Joseph was without space in their home for any more visitors. This is most likely because of their late arrival, and the home was already full from others who had arrived prior to them. According to what we know from the scriptural writings, Mary later gave birth to Jesus in something the world has depicted as a barn, or a stable. Let's look at the Scripture, and then we will explain this further.

> *And so it was, that, while they were there, the days were accomplished that she should be delivered. And she brought forth her firstborn son, and wrapped him in swaddling clothes, and*

> *laid him in a manger; because there was no room for them in the inn.* (Luke 2:6, 7)

As we look at this closer, we can see much written in these two verses. First, we see Mary wrapped Jesus in swaddling clothes. The basic definition for swaddling clothes is described as wrapping the baby in cloth tightly as a transition from the womb to the world. But, in this case, it's slightly different. The Greek word for "swaddling clothes" is *sparganoo*, which means "clothing (or to clothe) in strips of cloth."

As for the location of the birth of Jesus, we first have to understand that it wasn't necessarily in a stable or barn as we have been accustomed to hearing in the past. Looking at this a little closer, we can see that the accommodations of the holy family were not as we may have been taught. The structures in the ancient town of Bethlehem consisted of man-made buildings mixed with homes built into existing caves in the area. To understand the holy family's place of dwelling when Mary gave birth to Jesus, we need to first look at the Greek word for "*inn*," found in Luke 2;

> *...because there was no room for them in the inn.* (Luke 2:7)

The word is actually *kataluma (ref. 19)*, which is more understood as a guest room. It's important to understand what this actually would have been during the time period of the birth of Christ. Since this journey was due to a decree sent out among the entire Roman Empire, the couple would have most likely attempted to stay with relatives, as mentioned before. But with the journey slower for them because of Mary's pregnancy, they would have probably arrived much later than others in the family, and they were turned away from the guest room and sent to the lower end of the home where the animals most likely stayed at night. Considering that many homes were built into and above caves, the family would often bring the animals into the lower end of the home during the cooler nights. Soon after this arrival, Mary went into labor and gave a miraculous birth to Jesus, the Savior of the world, in a humble setting in a far-away town.

THE WISE MEN

Now, things will begin to get very interesting. Following the birth of Christ, deceit began to build in nearby Jerusalem. King Herod knew of the birth of a king but was strongly misinterpreting the magnitude of this King due to his pagan beliefs and lack of truly spiritual knowledge. Herod was more concerned with continuing his overall power along with his physical throne. He had been somewhat challenged by many, including his fear of his own children taking over his throne. So, his strong desire to find this newborn King became an infatuation for him. This theory is confirmed by what this power-hungry king sent out in the form of a decree later.

Before the forth-coming decree, there was another prophecy brewing by a group of men who were making their way to Jerusalem in search of the Christ child as well. We know these men in the biblical text as the *wise men* (Matthew 2:1). Little is known about these men, and the exact number of individuals that came seeking the newborn King has never been determined, only guessed upon. The Scriptures themselves are very limited to what we learn in them, but we can see many clues within the Word of God:

> *behold, there came wise men from the east to Jerusalem,* (Matthew 2:1b)

First, we see the wise men who arrived in Jerusalem came from the east. This is another placement of controversy. This one piece of Scripture has caused much of the world to believe the wise men, as mentioned only in Matthew, came from the Persian Empire. By the way, Matthew's writing of this account becomes more significant in what you will discover later. Unfortunately, the world is so deceived by only reading (or taught) the first thing they see, rather than looking at the Word of God as a full Scripture of the Lord. As you will see in these writings, it's most likely that the wise men arrived from the east because they had been traveling together in the Jordan River Valley (the Great Rift), which would be to the east of Jerusalem. Instead, we have jumped to the conclusion this one mention somehow tells us they were from the eastern part of the world. Instead, this is a simple mention of the direction they arrived in Jerusalem from in order to visit King Herod, because this was a familiar trading route through the mountainous terrain into the city. This fact is shown and proven when we continue to read the Scriptures. Knowing this, let's look at the reason of the wise men coming to Jerusalem:

> *Saying, where is he that is born King of the Jews? for we have seen his star in the east, and are come to worship him.* (Matthew 2:2)

This Scripture gives us a purpose, and more of a location of where these men actually came from. First, we see the purpose of the visit. They make a stop in Jerusalem, the capital city of the kingdom and the place where the king (Herod) reigned from. Not knowing the true purpose of the King they were pursuing, they were under the obvious assumption that everyone was coming to visit or worship this child which had been foreseen by them. This was a mistake on their behalf in the beginning, but would be later rectified by God's leading upon them.

The second clue we receive from the wise men is they came to Jerusalem initially from the west, moving north eastward, through the Jordan River Valley or outside trading routes, and turning back westward from the valley to arrive in the city. This fact is confirmed, once

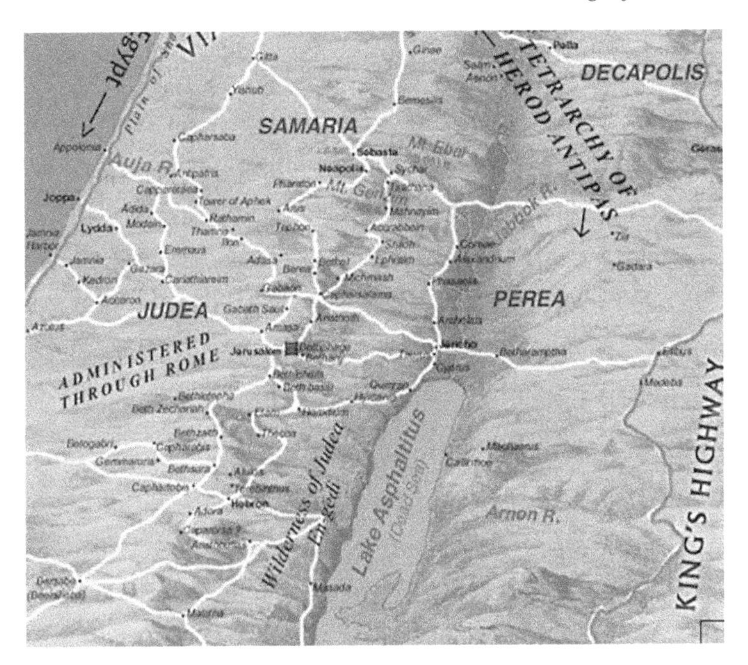

Map of merchants trading routes into Israel

again, by reading the Scriptures. Notice they mention they had seen the star <u>in the east</u>. If they came from the east, there is no possible way they could have seen the star in the east, and then ended up in Israel, or more precisely, Jerusalem. This false teaching over the years that the wise men came from the east is nothing more than either the lack of reading the Scripture as a whole, or to deceive from where these men would have most likely come from. As a matter of fact, in many new translations this Scripture has completely been altered to no longer reference the direction they saw the star in the sky. Some new translations will say they entered Jerusalem from the east but then change the reference of seeing the star in the eastern sky to various ways of simply saying they saw the star rise. It's a quick fix and a lack of understanding in order to avoid any questions of contradiction, when in reality the truth is very simple if we just follow the Scriptures. These men were guided eastwardly, from the west, (entering into Jerusalem from the eastern trading routes to be confirmed later) and explained to King Herod they were being led by the star in the east along their journey. Today, rather than learning, studying, and explaining the truth, the world wants to alter it to avoid any controversy and to be socially correct, but not historically correct. The Old Testament, Jesus' Bible, confirms where these men came from and we will thoroughly look at this closer later in this writing. That's why it's important to look at the Scriptures as a whole, and not to work in our own thoughts.

This Scripture again confirmed the wise men saw the star while coming from the west in verse 9:

> *lo, the star, which they saw in the east,* (Matthew 2:9)

This adds a whole new perspective on who these men were and how they were sent to find, and ultimately worship, the newborn King. This will all be confirmed later in these writings as you see this story continue to expand into a true-life event, as God's Word grows as one Word, and not a bunch of stories randomly placed into a book. It also verifies the Old Testament prophecy which many today won't even look at for confirmation, forgetting that this was Jesus' Bible of reference.

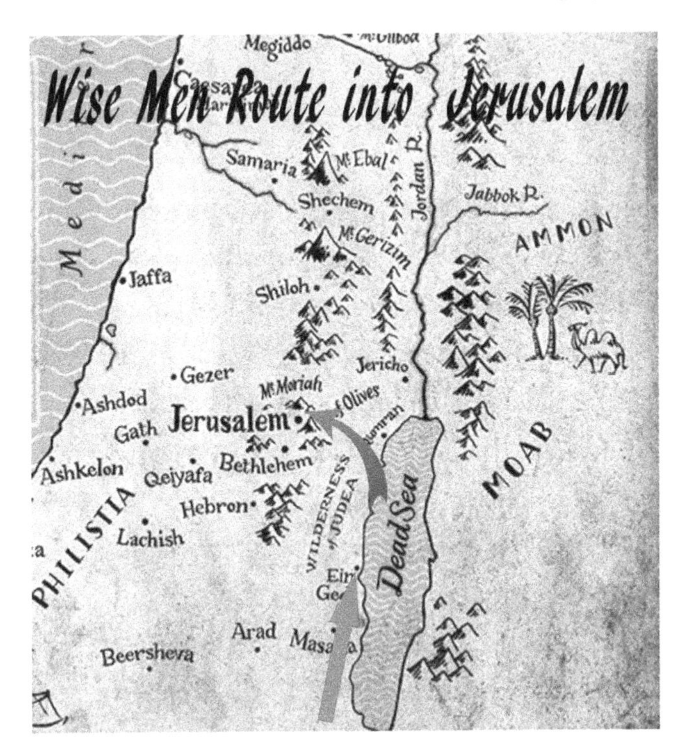

Map of wise men likely route through trading routes into Jerusalem from the east.

Continuing to the wise men's visit to Jerusalem, we see the inability of a jealous Herod to locate Jesus.

And when he had gathered all the chief priests and scribes of the people together, he demanded of them where Christ should be born. And they said unto him, In Bethlehem of Judaea: for thus it is written by the prophet, And thou Bethlehem, in the land of Juda, art not the least among the princes of Juda: for out of thee shall come a Governor, that shall rule my people Israel. (Matthew 2:4-6)

When the wise men came to Herod and described the child they sought, the king became fearful of losing his throne and called upon the priests and scribes (the scholars) to tell him of any prophecy concerning this matter. Once he heard this disturbing news, Herod attempted to convince the wise men that he wished to worship the child as well:

Then Herod, when he had privily called the wise men, inquired of them diligently what time the star appeared. And he sent them to Bethlehem, and said, Go and search diligently for the young child; and when ye have found him, bring me word again, that I may come and worship him also. When they had heard the king, they departed; and, lo, the star, which they saw in the east, went before them, till it came and stood over where the young child was. (Matthew 2:7-9)

Herod stooped so low that he lied to them to find Jesus, knowing the holy family would have sought somewhere to hide seeing the king's soldiers marching into Bethlehem from Jerusalem, which is just a few miles down the road. As these men left for Bethlehem, it appears as though they never thought Herod would mislead them. They simply went on their way to seek the child to whom they had journeyed so far to locate.

> *When they saw the star, they rejoiced with exceeding great joy. And when they were come into the house, they saw the young child with Mary his mother, and fell down, and worshipped him: and when they had opened their treasures, they presented unto him gifts; gold, and frankincense, and myrrh.* (Matthew 2:10, 11)

As we know, the wise men arrived in Bethlehem and were overwhelmed with joy as the star stopped above the site of the holy family. From this point we now realize that Jesus, Mary and Joseph were no longer in the *kataluma (ref. 19)*, but rather in a home. No longer is Jesus referred to as a baby, but a young child, which gives us a clue to the age of Jesus when they finally arrived. Most likely Jesus was near the age of two. We determine this because of the translation of the Scriptures reference to a *young child*, and to the decree of Herod to come. Again, overwhelmed, these men fall to their knees and worship Jesus. This is a site that you can only imagine, as the power of the presence of the Lord is revealed to these men. They were not only overwhelmed with joy, but they were brought to their knees with the radiance of what they saw.

As the wise men accomplished their purpose of delivering the gifts of gold, frankincense, and myrrh to Jesus, they departed away from Bethlehem, defying the king's request to return and tell him where the child was located.

> *And being warned of God in a dream that they should not return to Herod, they departed into their own country another way.* (Matthew 2:12)

God stepped in at this point, and through a dream, led them away, warning them of Herod's deception. The wise men departed and returned to their own countries, where I'm sure they would have shared what they had seen. As for King Herod, the fear and anger had festered within him as the wise men didn't heed to his deceptive request, and it was at this point that he issued the order of decree.

> *Then Herod, when he saw that he was mocked of the wise men, was exceeding wroth, and sent forth, and slew all the children that were in Bethlehem, and in all the coasts thereof, from two years old and under, according to the time which he had diligently inquired of the wise men.* (Matthew 2:16)

It was this Scripture that shows us the reality that a great deal of time had passed since the birth of Christ and when the wise men arrived. It was also during this time that the Bible tells us Joseph had a visit from a Holy being;

> *And when they were departed, behold, the angel of the Lord appeareth to Joseph in a dream, saying, Arise, and take the young child and his*

> *mother, and flee into Egypt, and be thou there until I bring thee word: for Herod will seek the young child to destroy him. When he arose, he took the young child and his mother by night, and departed into Egypt...* (Matthew 2:13-14)

This was, without a doubt, a frightening call upon Joseph and Mary, to say the least. Remember, this was a couple who had just traveled a rugged journey from Nazareth to Bethlehem upon the order of Caesar Augustus. Now, they are told to continue onto a longer trek into a foreign land that would be completely different in all ways to what they had grown accustomed to. As I wrote in my book *Expedition Ark of the Covenant (ref. 20)*, this was the beginning of a journey that would prepare this child for the ultimate sacrifice for the Salvation offered to all of humankind.

THE MATTHEW FORETELLING

What's more astonishing about this Scripture discussing the wise men in the Bible is that Matthew was the only disciple who wrote about this event. Nowhere else is it written about the wise men's visit to see Jesus, and it's only Matthew who gives great detail to their coming, their reason behind the visit, the description of Herod's horrible decree, and the beginning of ten missing years in the life of Jesus. Matthew also gives us a closer look at where these men came from and to the overwhelming joy that was received when they saw and presented the holy gifts to the Messiah.

The holy family was about to embark on a journey that would be the beginning of Jesus' ministry on the world, as we shared through the discovery of the missing manuscripts of the purpose of their trek into Egypt and beyond in my book, *Expedition Ark of the Covenant (ref. 20)*. But this revealing text also gives us a glimpse into an acquaintance that Matthew had later in his life, and one that will bring the Bible together for you by the end of these writings.

God's Word is truth, and biblical history is truth if it aligns with the Bible. That's where this adventure begins, so let's move forward in sharing an incredible story that has been lost for centuries. Clues have been laid out in front of us by those closer to the years of the original texts, and those will help bring the Bible to life in an even greater way than ever before. But understand that your faith can only grow if you're willing to step out of the box of tradition (which you have seen in the wise men's reference to where they truly came from, and where tradition has always said they were from) and step into a whole world of faith through the Scriptures and through the ancient writings that the men who God entrusted to write His Word were referencing. By the way, later we will offer even more biblical proof of where these men originated from and why this is so important to the story of Matthew. God is an amazing God, and when we're willing to step back and see His Word live in truth and not in tradition, then we are able to see God in all things within your own life. And I mean in every part of your life.

Chapter 3

Ancient Historical Manuscripts

Early church leaders (top row l to r): Clement of Rome, Papias, Polycarp, Jerome, Eusebias; (bottom row l to r): John Foxe, Jacobus de Voragine, William Tyndale, John Wycliffe, Martin Luther

THE TRUTH OF THE Scriptures can be shouted when each and every archeological discovery is made, and when each story in the Bible is brought to life. From front to back, Genesis to Revelation, God's Word is spoken to us as truth daily in our lives. As much as the secular world, and sometimes even those claiming to have faith, would attempt to prove the Word is not truth, the archeological evidence over the ages more now than ever proves the truth is the Bible. But what about before the Bible? The Word of God we know today is filled with history, prophecy, revelation, guide to salvation, and training. It gives us the history of God's creation and through the ages of the early foundations, along with law of the Lord laid out before us. It gives us the prophecy of those God spoke through to share with us the guidance for the events and times to come. It goes further to reveal what will be needed for us to live in the fullness of life through His Son, Jesus. And finally, the Bible gives us the necessary training to move forward into a fruitful and truly blessed life. But there was much still happening in biblical times. The Bible was being lived and was not being written down during that time. But those of the Scriptures refer many times to books that had been written to give reference to what has already happened in time. The same is true today when we reference the likes of Abraham Lincoln, Julius Caesar, or the Revolutionary War. We turn to the books written during the period to give us the history of those people or the events we are referencing.

One might say that there is no way the Bible can be looked upon as a guide book to anything, while others will claim that confirmations to lead one's path to the next level are simply coincidence, not any divine leading. Pastors across the country have asked me many times to go further in writing a book explaining how to see God's guidance within every step in your life. I truly believe God is in everything, but we are so close-minded to see Him in those things. Over the last several years of research, exploration, and study to confirm these, and many other findings, there have been unbelievable confirmations that have occurred to keep me moving on this path. Coincidence? I don't believe so. The confirmations I have received through people and events are in no way coincidence. They are the bread crumbs of God leading along the path to complete a task He has set before me.

Duart Castle, Scotland

Let me begin by stepping back with one of my ancestors. The Rankin Family came from Scotland and were part of the Clan MacLean and served at Duart Castle in the 14th Century. The family had many who were martyred for their faith, and a few were able to escape to seek a new life in the new world. Thomas Rankin crossed the ocean and ended up in Pennsylvania, making his way to Eastern Tennessee, and settling in a place they called Dumplin Valley, alongside the area founded by his friends, the Crocketts and the Findleys. Yes, that is David Crockett's family, along with his wife's family, the Findleys. Thomas and his wife gave birth to Richard, who fought in the Revolutionary War, alongside his friend John Crockett (David's father) at the Battle of Kings Mountain. Richard eventually married, and later his wife gave birth to John Rankin. This is the man I would like to focus on for a few minutes. John attended Bible seminary and became very outspoken on his stance that no man should be enslaved. This was very unpopular in East Tennessee, thus John and his wife moved away, first through Carlisle, Kentucky, and then eventually settling in Ripley, Ohio along the banks of the Ohio River, just east of Cincinnati. After a few years there, he built a small brick home in 1829 high atop a 540-foot hill overlooking the town, the river, and Kentucky. Reverend Rankin was becoming a well-known abolitionist and was the lead pastor at a Presbyterian church in Ripley. His reputation for his support of abolishing slavery was well known as hecklers would try to prompt him while bounty hunters, and slave owners, placed a price on his head well into the thousands of dollars. But Rankin built a force to be reckoned with atop that hill with his wife and their nine children at the time (13 children as years moved on), along with many from his family that he took in. Those seeking freedom would arrive at the banks of the river in Kentucky and look for the light that Rankin left burning on the hill. If those seeking freedom were able to make it across the river, Rankin's sons, or sometimes with the help of several others, would assist them to climb the steep wooden stairs to the reverend's house to be hidden until the sons could move them onto the next stop to freedom.

Reverend John Rankin

Many people were influenced by John Rankin and his no-nonsense attitude toward slavery, including William Lloyd Garrison, Frederick Douglas (who was beginning to fight for the abolitionists movement alongside William Lloyd Garrison), Henry Ward Beecher, and author Harriet Beecher Stowe who based her book *Uncle Tom's Cabin (ref. 21a)* from two of those individuals that Rankin helped lead to freedom. It was Harriet's brother, Henry Ward Beecher - a popular clergyman, speaker, and diplomat who, when asked after the end of the Civil War, "Who abolished slavery?" He simply answered, "Reverend John Rankin and his sons did." Rankin, to his dying breath, continued to speak out against anyone who would attempt to own another man or woman's life. He and his family were responsible for helping nearly 3,000 men and women seeking the life of freedom that the Lord had promised. As I began to study Rankin's life further, it also opened me up to more history of the family involving John's sons, including influence on the election of Abraham Lincoln in 1860, Ulysses S. Grant's first officer appointment in the army, involvement in the earliest ice rinks in America, and refrigeration as we know it today.

In the summer of 2015, Sherri and I were traveling into Tennessee for a series of speaking engagements, and we took an afternoon and drove along a winding road next to Douglas Lake just outside of Sevierville. We came to a small quaint little town called Dandridge where I saw a parking spot at the side of the road. It was a cool little place, like time had somewhat stopped minus the new cars everywhere. As I pulled into the parking slot along the sidewalk, Sherri said, "Look at that old cemetery over there." Directly across the road was an old Revolutionary War cemetery with a bent wrought iron fence around the front. As we walked closer, there was a pillar monument, and then around it was a series of graves strewn throughout. Many were so old they were unrecognizable but it was the stone pillar that stopped us in our tracks. The stone column had a plaque attached that said this cemetery was "In memory of the Revolutionary Soldiers Buried Here." As we read the five names on the plaque, we stood and stared as the names of Richard Rankin and Samuel Rankin were among those listed. We immediately went over to the courthouse where they directed us to the archival department. Here we found two wonderful volunteers wanting to know how they could help us. We inquired if they had anything on the Rankins mentioned in the cemetery, and they asked us why we were interested in that. I said, "I'm a Rankin and we wanted to see if they were any relation." As they spoke, I was taken aback to find out that the Richard on the pillar was John Rankin's father, and this was where he was born. They began to pull books and books from the shelves of family history, including those of my direct relations of that time period. They said there's a whole cemetery full of the Rankins in Dumplin Valley, and the upper part of this lake is called Rankin Bottoms. We sat there overwhelmed.

But the fun was about to begin as it all confirmed we were on the right path. They slid us a copy of the original letters that John Rankin had written to his brother Thomas who had purchased slaves while living in Virginia. John was somewhat chastising his brother, but in a way to share truth through the Scriptures with him. These, now known as *Letters on American Slavery*, had become standard reading for abolitionists from all over after being published by William Lloyd Garrison *(ref. 22)*.

As I opened up the letters, it didn't take long before I sat there in tears as God began to confirm our trek. Throughout John's writings he consistently spoke of the Ethiopian people (referring to the large central part of the continent of Africa at the time before countries had been divided and named) and the importance they were in the Bible, and that we all came from the same blood.

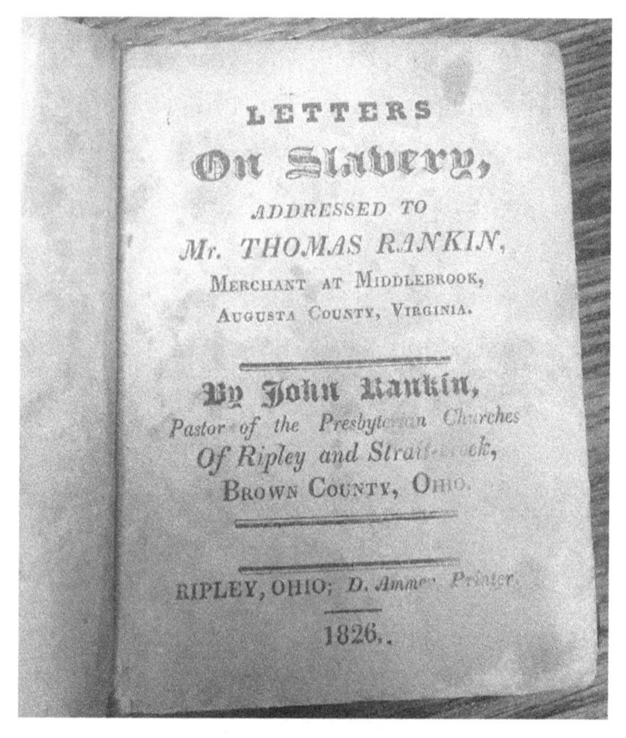
LETTERS
On Slavery,
ADDRESSED TO
Mr. THOMAS RANKIN,
MERCHANT AT MIDDLEBROOK,
AUGUSTA COUNTY, VIRGINIA.

By John Rankin,
Pastor of the Presbyt[illegible]n Churches
Of Ripley and Strat[illegible]ck,
BROWN COUNTY, OHIO.

RIPLEY, OHIO; D. Amm[illegible] Printer.
1826.

Reverend John Rankin's 'Letters on American Slavery'

In the first letter Rankin explains about the African skin as the world had been somewhat brainwashed that the darkened or "*black*" skin of the Africans was a curse for slavery. He says, "*. . . let me invite your attention to the book of inspiration; there you will find that the blackness of the African is not the horrible mark of Cain, <u>nor</u> the direful effects of Noah's curse, but the mark of a scorching sun. Look not upon me, because I am black, because the sun hath looked upon me: my mother's children were angry with me; they made me the keeper of the vineyards* (Canticles 1:6) (*ref. 22*). First, let me explain what the Canticles are. Early church leaders from the initial years of translating the Scriptures, and continuing as late as the latter

1800s, used these as additional hymns or even psalms. It was later found these were originally part of the biblical Psalms as found in the Dead Sea Scrolls.

Although I had done a great deal of study on the historical manuscripts, many of which are mentioned by name in the Bible or quoted, I had not fully applied them as research materials in locating lost biblical locations and finds. The men who wrote the biblical text didn't have a Bible to reference because they were in the process of living it and writing it. Thus, they referenced historical books of the time period, which were later referenced by the Jews, early church historians, and Bible translators. This was a new open door that had to be gone through because of the housing of many ancient texts that I had encountered in hidden locations in Ethiopia.

A final thought on Reverend John Rankin, whom I respect greatly for his desire to serve the Lord, no matter the criticism. My wife and I were traveling through Kansas, speaking in churches throughout the area with our dear friends Bev and Wayne Regehr. Bev is a key part of our work, and she has been there since the very beginnings of the work and discoveries in Ethiopia. Bev's brother-in-law, Roger, came to me and said that he had this strong feeling that John Rankin (whom he had done some research on) not only was writing these letters to his brother and to the people of that time, but he was also writing me a letter nearly 200 years ago. The letter was to help me along my journey with these ancient texts and to confirm that we are serving in the right place by sharing with the Ethiopian and African people, but also to the world of their importance.

This was an amazing eye opener for me to pursue further this God-lead journey and the world changing information that has been opened in my own path.

HISTORICAL MANUSCRIPTS AND THE BIBLE

Ever since my mother gave me a small pocket Bible when I was a little boy, the contents of the Scriptures have been a thing of astonishment to me. The inspired Word of God, actually breathed by Him into the hearts and minds of men to record, was given to us as direction, sometimes the correction for our actions, and the simplistic link to being accepted by Christ. Many people claim that salvation through Jesus is a complicated happening, and once you receive Him into your life there's a process to keep that desire going. My answer to that: once truly received into your life and the Holy Spirit is dwelling there, you can't help but want to serve Him and serve others in His name. But knowing that each word in the Scriptures was given by God Himself to those who transcribed it is a humbling reminder of His love for us all.

> *All scripture is given by inspiration of God, and is profitable for doctrine, for reproof, for correction, for instruction in righteousness: That the man of God may be perfect, thoroughly furnished unto all good works.* (2 Timothy 3:16-17)

That's why anything dealing with the Scriptures and continuing with the discoveries about the historical past needs to be dealt with extreme care.

Our Bible was given to us by God, and beyond that it has been man who has decided in future translations what should be included in it through prayer and even some turmoil. But in the process, we must be cautious what we do because of what John was told by the Lord in Revelation 22:18-19:

> *For I testify unto every man that heareth the words of the prophecy of this book, If any man shall add unto these things, God shall add unto him the plagues that are written in this book: And if any man shall take away from the words of the book of this prophecy, God shall take away his part out of the book of life, and out of the holy city, and from the things which are written in this book.* (Revelation 22:18-19)

There is no question that we have seen many attempts to add to the writings of the Bible and even take away pieces of the Scriptures by some of the translations that have been presented for us to consider over the years. Some people are staunchly set on one Bible translation or the other. I, for example, use two versions that are closely related, but I will not degrade anyone who chooses another version. However, I do urge everyone to be aware if things have been added, changed, removed or simply moved to a footnote at the bottom. Today, many pastors even use various versions of the Bible to make a point in their messages. For myself, I'm not one in favor of that, just because I have witnessed confusion by many trying to follow along, but I am not going to criticize for doing it, unless it is being done to intentionally confuse in order to push an agenda.

When it comes to the ancient historical writings, they complement the original biblical text, and in no way add to the Scriptures but in most cases continue in the lives or events in the Word of God. Many of these writings are unknown in their origins and may have been written originally by the same men who gave us many of the writings in the Bible but were not to be included in the Holy Scriptures. For example, the Book of Jubilees (which has long been used by Jewish leaders, and even recognized by early church fathers such as Epiphanius, Justin Martyr, Origen, Diodorus of Tarsus and others) was said within its pages to have been originally written by Moses continuing the true story of Adam and Eve and beyond, with additional translations added later. Over and over again in the last 100 or more years, we have been given a great privilege to see the revelation of God's secrets and mysteries He mentions in the Bible: documents, items, artifacts, locations, and sites that have been sitting dormant for thousands of years but have been newly revealed to the world. Many are still in question because it hasn't met the criteria that modern-day man has set-forth as criteria, but it may match more closely to the biblical description.

These revelations have been happening so quickly and so often that the world has literally become numb from seeing or recognizing these finds. The mainstream news no longer reports these findings, when in the days of the past it would have been on the headlines. This should show us that God is being put on the back page as well in our lives. I mean, when you really think about it, who is it that causes much of the confusion when it comes to the findings of the artifacts, scrolls, writings, and locations? It's usually from the Christian world, whether it be theologians who try to disprove anything new mostly from their lack of discernment, or those church officials who claim that nothing else ever happened that's not mentioned in the Bible. We all know that's not true, and for those who are constantly trying to disprove any new find, well, it's done nothing but cause confusion to those seeking the truth or answers to questions. I can say that from experience. I was that person who seemed to never get a solid answer other than, "The Lord works in mysterious ways", or "If it ain't in the Bible then it didn't happen." How ludicrous for someone to preach or think that!

On the other hand, we need also to cross-check everything. There are a great deal of forgeries and false claims out there in the world as well. But one claim that I will stand up for is that God didn't say that you had to have multiple Ph.D.'s to be able to locate and find the secrets of the Bible. How do I know? Well, I've read the Bible and seen that simple, everyday people like you and I can also be a part of what God wants to show us in His Holy Word or even halfway across the world in His lands if we're willing to be led and shown. God is revealing, and we truly need to open our eyes to the possibilities rather than dismiss them. If the Lord is blessing, then maybe He truly is behind it after all. Always remember, as I mentioned earlier, that our world looks at many things in the Bible as impossible, but in God's world we need to let God out of the little box many of us try to keep Him in and see that ALL things are possible.

HOW WE GOT OUR BIBLE TODAY

This topic is always much in debate: *How did we get our Bible today*? As mentioned before, we know that the original Scriptures were inspired or breathed by God into many writers. Just as I woke up one morning with a dream about the Ark of the Covenant that has now changed my life, God did the same thing through the men who wrote the Bible. It was many years ago when I had a dream about the Ark of the Covenant which became reality just over a year later when I discovered the footholds of the Ark in Ethiopia simply as part of a team expedition. Really, the same thing happened with many of the men who wrote the Holy Word. It goes without saying the way it has been read and translated over the years has caused much confusion in the world. The Bible is revealed to you. It's revealed in a time when you need to know, and in a context, that you need it for. But there are those who will take the Bible and literally try to create their own word from it, even to the point of nearly rephrasing it to fit a need.

God gave us the Bible in various books and epistles (or letters) to inspire, train, share, and give hope along with allowing us to receive salvation through the Lord Jesus Christ through our faith in His grace. The first five books were written by Moses, inspired by God, sharing the beginning and then beyond with what had taken place through his own time. We see prophets sharing what God had told them was to come. The gospels were written by men whom God inspired to tell of the story of His Son, and then much of the New Testament was given to Paul for the continued purification of the saints. And finally, the book of Revelation was to give us a glimpse into what was to come. You could almost say that John, the writer of the Revelation, went from an apostle to a prophet with his works in this book. After the Word was given, it was passed down through the ages, copied from scroll to scroll and from book to book. Much of it was passed down verbally and then later written, while to others it was given from place to place by scribes who spent months copying the writings from one handed down to them.

Many have understood the Bible wasn't even in writing in its earliest years but verbally passed from teachers or priests to the next. The most common belief is that the Bible never really made it to written form until 1450 to 1350 B.C. or somewhere around that time. Then there was a time of nothing added to the writings until the New Testament when much was given in a very short period of time. Even Paul, who wrote the largest portion of the New Testament, gave us a glance into how this was done in 1 Corinthians 15:3:

> *For I delivered unto you first of all that which I also received . . .*

It was also known that for up to at least sixty years after the time of Jesus, there was more oral passing of the Word and knowledge of God to others than there was in written form. One great example of this can be seen when you visit the churches and monasteries in Ethiopia. You can physically see the passages of the past, both biblical and historical, in hand written form that was passed down orally through the ages.

Looking back on those who gave us the Scriptures in the forms that we see somewhat today are people such as Origen, who spent most of his life in Alexandria studying and translating Scripture in the 3rd century. Pope Damascus was the one who inspired Jerome in the 4th century to begin his work in translating the Bible into Latin thus giving us the original Latin Vulgate. In the 5th century, Caedmon of England began singing the Scriptures in English in the Whitby Abbey after feeling challenged by God to sing them. There were many other translations of different books into various languages, but it was in January of 1384 that John Wycliffe was involved in the translation of the Bible from the Vulgate into English (*ref. 23*). Martin Luther translated the Scriptures into the German language as part of his battle against the corruption in the church in 1522. And then through his works from 1525 through 1529, William Tyndale took the Hebrew and Greek and translated the Bible into English. He's known as the "the Father of the English Bible" as it is claimed that much of the *King James Version Bible* is based on the work of Tyndale. His influence, however, came from the study of Luther's translation while hiding out in Germany to do his work away from persecution. Tyndale's work was printed on the Guttenberg Press, and he suffered much for his work; he was eventually burned at the stake in 1536 by order of the Bishop of London (*ref. 24*).

In 1611, after numerous other finished translations had been inspired by men to publish, such as the Geneva Bible and the Great Bible, the *King James Version* of the Bible was completed by the order of King James I of England (*ref. 25*). During this process of translation by various scholars of the Christian denominations of the time, there was much prayer, bitterness, compromise, and eventually agreement to complete this work (*ref. 26*). Since the 1611 *King James Version Bible* was released in print by Robert Barker, and with its revisions over the years, especially into the standard text of 1769, there have been dozens upon dozens of translations for various reasons such as denominations, modern language, or to attempt to make the Bible more interesting, or so they say. Unfortunately, many of these have made the Bible a book of fables or have watered down the Word of God to make it nearly unrecognizable. Or some were simply to fit the desires of a denomination and what their beliefs needed to be. There are good works out there, but one must be cautious to the points I have mentioned. And no matter what, when man first touched the Word of God, it has become more tainted each time it is translated, which many times compromises the wording, all in the claim to make it more understandable to the general public. As many have said before me, when God wants to reveal something to you, He will clearly show you. Too many translations today, and there are hundreds of them, have altered, changed, manipulated, and even removed Scriptures from the general text to supposedly help us understand God's Word more. Really?

When it comes to the Bible and what we see in it, has been a somewhat rocky process. After the Scriptures were compiled into an accepted (canonical) form, there was much controversy on what would be included and what was not. For example, the book of Esther was not included in early lists of the canonical writings. As a matter of fact, Esther is the only book not found in all of the Old Testament books discovered in the Dead Sea Scrolls, which has many questioning whether this book should have been originally included in the canonical Bible and how it got there in the first place. Many other books in the Bible today have been subject to protest, including the book of Revelation and the book of James (*ref. 27*). But yet there were many other books that were widely accepted as canonical books in years prior to the 20th century that were challenged and not included in the Bible today as decided by generations expanding further away from the translations of the original works.

DEAD SEA SCROLLS

Now let's look at one of the most favored, while at the same time most controversial of the discoveries of the texts of the Bible. Without any doubt the discovery of the Dead Sea Scrolls in 1947 in a series of caves in Qumran (located near the Dead Sea) has most definitely confirmed that the books of our Bible today are accurate for the most part. There are differences, but for the most part they are close. There are some questions that come to mind when looking at the Dead Sea Scrolls, though.

The scrolls were accumulated by a group of people called the Essenes, who were much like the Pharisees and Sadducees of the day but more secluded and grounded in their beliefs and preservation of the Word of God in the Old Testament. They are believed to have been guardians of the scrolls, and hosted an even deeper and more mysterious belief to preserving this incredible library. Historian and former Roman general Flavius Josephus says in his works *The Jewish War: Book II* (Chap. 8), "*Although Judeans by ancestry, they are even more mutually affectionate than the others. Whereas these men shun the pleasures as vice, they consider self-control and not succumbing to the passion's virtue . . . And whereas everything spoken by them is more forceful than an oath, swearing itself they avoid, considering it worse than the false oath; for they declare to be already degraded one who is unworthy of belief without God. They are extraordinarily keen about the compositions of the ancients, selecting especially those [oriented] toward the benefit of soul and body*" (*ref. 28*).

The fact still remains: the Essenes were the keepers of this incredible history, and right alongside the books that we recognize as canonical, or biblical, were many other writings they studied as historical documents to help confirm what they were translating in the Bible. Not until well after these original keepers of the ancient texts had taken their works to other areas and were gone did man decide that some of these texts were just too big in their context of events to be included. The works found in the Dead Sea caves have been long considered the missing link to giving the Scriptures life beyond the Scriptures, but there are many questions that are raised about the scrolls.

One of those questions is this: If we are going to look at the accuracy of the scrolls as being authentic, then what about many of the other books found right alongside the canonical books that the Essenes were preserving, studying, and using for confirmations to what they were writing? Yes, I agree, these historical writings should not be considered as biblical text, but if many of these books were used as historical works by the Essenes, whom we have put so much trust into their duplication and study of the Scriptures as preserved in the Dead Sea Scrolls, then maybe we need to look further into the possibilities that we should also accept them as maybe not canonical but as historical documents to the events in which the Bible only touches upon. It should also be a confirmation of the ancient writings of the Ethiopians which have had these works in full versions in their possession for centuries.

Books such as the Prophecies by Ezekiel, Jeremiah, and Daniel are found in Dead Sea but not in the Bible. Also, the long talked about missing Psalms of King David and never before seen Psalms of Joshua are found in scrolls. And then you can find the last words of figures such as Joseph, Judah, Levi, and Moses' father Amram written in some of the scrolls along with them (*ref. 29*). Within these caves were also found fragments containing previously unknown scrolls of Enoch, Noah, Abraham and others. Alongside of biblical text were located several writings of the Book of Jubilees and the Book of Enoch, which until that time were only found in full form in ancient Ethiopian writings. I have held nearly two-thousand-year-old copies of these books in my hands that are hidden away in remote locations, away from the open eyes of the world, which for me has been breathtaking. Without a doubt, we must be able to look beyond our comfort zones and consider the possibility that the other books alongside the canonical scrolls may have merit to unraveling many of the mysteries that the Bible talks about. But we also must consider that the answers may not be resolved the way we want them to be.

NON-CANONICAL BOOKS MENTIONED IN SCRIPTURE

There has been much controversy about the mentioning of other books in the Bible. There has been much debate, much argument, and much deceit in this process. Many have argued that no other books were referenced in the Bible except for those within the Scriptures themselves. Others will claim that since the wording used within the Scriptures of these so-called other books aren't quoted word for word that they couldn't possibly have been from another writing.

Let's take a look at this in two parts. First: since the writers of the Bible, especially those of the Old Testament, had no other references to the Scriptures then they, of course, would have to compare some of their experiences to other books that had been written. For example, we have references by Joshua to the Book of Jasher (Book of the Upright One or the Book of the Just) (Joshua 10:13):

> *And the sun stood still, and the moon stayed, until the people had avenged themselves upon their enemies. Is not this written in the book of Jasher? So the sun stood still in the midst of heaven, and hasted not to go down about a whole day.*

and also (II Samuel 1:18):

> *Also he bade them teach the children of Judah the use of the bow: behold, it is written in the book of Jasher.*

Here are clear clues that these books were used regardless of the way they appear today. Sure, they have changed over time just as our Bible has in some ways. But they were still referenced. Other references include The Book of the Wars of the Lord in Numbers 21:14; The Chronicles of the Kings of Judah and the Chronicles of the Kings of Israel are found in 1 Kings 14:19, 29, 16:20; The Book of Shemaiah and Iddo the Seer are seen in 2 Chronicles 9:29, 12:15, 13:22 and numerous others including The Acts of Solomon, Annals of King David, Book of Samuel the Seer, Book of Nathan, Book of Gad, and others (*ref. 29*).

As to the second half about the other books is the fact that because the quotes in the Bible are not the exact wording, then there is no way that they can be referencing these books. Boy, have I heard that one enough. Let me say this: Have you then looked at the four gospels in the Bible? None of them use the exact wording that even Jesus spoke. They all record similar events, but many have questioned the authenticity of the Bible because they aren't given in the exact timing as the other books record, such as those who went to the tomb of Jesus after the resurrection. In defense of the writers of the gospels, for example, those that saw the event wrote it differently depending on the time of the day their part or description took place. Even direct quotes of Jesus are varied because when they went back to write it down, they did so remembering in a different way, but they all complement the other. When I quote Scripture without looking at it in the Bible, sometimes I don't get it word for word no matter how hard I try. For people to say that these quotes couldn't have been from other books is just another excuse and a weak one at that.

Let it also be known that scholars, for the most part, were very much against many of these writings in reference until they surfaced in Hebrew and Aramaic in the Dead Sea Scrolls. Prior to that, as I mentioned, the only known full form of these books were found in the Ethiopian form but were denied because of that. Later, many began to rethink all of this after their discovery in the caves of Qumran.

SECRETS TO BE REVEALED

There is no doubt that the Bible tells us that there will be more to come. Its constant reference to the revelation of secrets and mysteries gives us hope that even more evidence to prove the Scriptures will be located. That's why we are seeing such a rush of archeological finds and discoveries than in any other time in history. We learn in Matthew 10:26:

> *. . . for there is nothing covered, that shall not be revealed; and hid, that shall not be known.*

and then in Luke 8:17:

> *For nothing is secret, that shall not be made manifest; neither any thing hid, that shall not be known and come abroad.*

It's exciting to know that the Lord will reveal in His time. But with the abundance of evidence now coming forward, we must heed to the fact that His time is now for some of these mysteries (Daniel 2:28):

> *But there is a God in heaven that revealeth secrets.*

Some of these came true when we made our first finds several years ago which have led to this process and discovery. It was John revealing one of those secrets that helped open the door for us. (John 21:25):

> *And there are also many other things which Jesus did, the which, if they should be written*

> *every one, I suppose that even the world itself could not contain the books that should be written. Amen.* (John 21:25)

There is much to be found, identified, discovered, and sought after if only we will be open to see that God is there and that He can, will, and has done great things. Many times I ask the question, "Why us?" What's even more amazing is when we have pastors and others say that "You are just like the unlikely that we find in the Bible. God has prepared you for many years to be doing this but you didn't know it." It was during one of these times when we had this deep conversation among a few of our team members in our ministry. As we passed through Dayton, Ohio, we stopped at a Wright Brothers site to take a break from our travels to see the story of these young men who changed the world through flight by discovering long hidden secrets to flying by observing birds. I turned a corner in the exhibit and right there in front of me was a wall with a huge picture of Orville Wright sitting on the ground working on his flying machine beside a building. But then the quote on the wall left me in tears just moments after our conversation: Why us? The quote said: "*Isn't it astonishing that all these secrets have been preserved for so many years just so that we could discover them*" (*Orville Wright*).

The secrets of God's Word are being discovered. They are being revealed to the well-learned and trained, but they are also being discovered by the most unlikely as well. That's you and me. Why? Well, maybe, because it's not just about how much you know, but how much you're willing to believe. And those willing to believe will have the faith of a child, not to question, but to follow and trust in what God is revealing. Again, why? Because many times God's revelations of supernatural happenings are far beyond the realm of belief for those who need definitive proof away from the logic of God's Word.

BOOKS OF QUESTION: REVEALING THE TRUTH THROUGH NEW FINDS

Without any doubt, I truly believe the Lord is showing more than ever, because more people than ever are trusting in their own discernment through God than in what others are saying. When you think about it, people are so tired of the drama, heartaches, fussing, and gossip that you get on social media, and are more willing to step out of the box to find the Lord can really open your eyes if you truly desire to see. When I used to preach in a church in the United States, I would always tell the congregation that they should never trust in what I'm saying . . . they should always open the Bible and research it on their own. Not that I would lead them astray; it's just that you should always see what God is going to reveal to you through His Word.

In my studies and discussions with well-learned individuals and then my actual hands-on explorations, I have been able to see something very similar to what I used to tell the congregation. For centuries, men have tried to suppress the church from learning and knowing that there may be other books that were written in history that were possibly inspired either by God or by those who witnessed events take place that may be lingering out there over time. Just as the early church tradition was, many may have passed the history of the events down through generations, just as the Holy Scriptures were, and then somewhere in time someone decided to record them on papyrus, goat skins, or even on stone. Then after my own investigations into some of these books, and watching to see how hard some try to hide or keep others from seeing them, I really felt that there may be more to the story than I first thought.

You see, many of the books I'm mentioning here were quoted in the Bible in various contexts or other names, and they were welcomed by the early church fathers as historical accounts, or on the other extreme, were removed or even attempted to be destroyed to protect an agenda that was taking place over time. Then I was truly amazed when I began to speak of some of these writings to some very well-versed men of God who pulled me aside and said, "I believe in these, too, but there's not many that you can talk about them with because of tradition." Again, not that they should be included in the Bible, but that they should be examined for their historical content to give a further possible history of the events in the Scriptures.

But the most astonishing factor in all of this is that when I investigated these writings further and matched them with the Bible, and then visited the locations that many of these sources led to, that's when things began to heat up and the locations became very visible. Many put so much effort and trust in Josephus for his historical writings but disregard anything else written over time that they have missed the possibilities that

God may have allowed. Unfortunately, those trying to keep you from seeing these texts are most likely the same ones speaking from one side of their mouth and saying something else out of the other.

Over time, many of these stories of early Bible events continued to be passed down through generations, and that's why those closer to the times when the Bible was written even accepted these writings and knew of the events and locations we have now discovered. While others have discounted them, even though the Ethiopian full texts that were recorded for generations, they were all shocked when versions of these finds were discovered in the Qumran Caves. The Book of Enoch is a great example of this. The Ethiopians have had hidden away for centuries the full text of the Book of Enoch, and some of these copies are very old. By the western world it was dismissed because it was only found in the Ethiopian ancient language of Ge'ez. But then during the discovery of the Dead Sea Scrolls it was found in parts in Hebrew, Aramaic and even in Greek. Many have claimed that just because the Book of Enoch has had other sections added to it over time, such as works related to astrology and other corrupt works, that it could have never been written by the well-versed Enoch in the Bible, a man favored by God. That same theory goes right along with what I had said before; the stories have been passed down as historical content, and then through time others have added to them, which in this case was true when an Egyptian had a hand in some of the early translations. You have to be able to discern and then let the Lord guide you to truth, something that many of the theological world have lost because they have buried their heads in books rather than seeing the world from reality and true faith.

The Book of Adam (Book of Adam and Eve), Cave of Treasures (Apocalypse of Moses), Book of the Bees are very much the same, as is the Book of Jubilees. Even though these books have appeared in various forms and texts, many were also part of the Jewish Talmud. It is believed that the Book of Adam was written by an Egyptian, but the case that he might have been the first to actually record it is most likely the actual truth. Again, the Ethiopians have had these in full text hidden away for centuries, but the attempt to remove these biblical people from any importance has proven to have put the joke on the rest of the world. It is even believed that the below references in 2 Corinthians are attributed directly to quotes in the Book of Adam & Eve:

> *. . . for Satan himself is transformed into an angel of light*. (11:14)

> *. . . God knoweth;) such an one caught up to the third heaven*. (12:2)

And then there is The Assumption of Moses referenced in Jude 9:

> *Yet Michael the archangel, when contending with the devil he disputed about the body of Moses*.

And the Book of Enoch is said to be referenced in several locations, but none is as visible as in Jude 14:

> *And Enoch also, the seventh from Adam, prophesied of these, saying, Behold, the Lord cometh with ten thousands of his saints.* (*ref. 30*)

One other note to the Book of Enoch: it was found in numerous caves in Qumran amounting to copies more plentiful than those of many of the canonical Bible books.

The Ethiopians have held in their full form nearly all of the books found in Qumran, but the more educated have decided that you shouldn't see them, even though early church fathers, translators, and educated men have accepted these writings. One critic of the Book of Adam said, "*This is we believe, the greatest literary discovery that the world has known. Its effect upon contemporary thought in molding the judgment of the future generations is of incalculable value.*" And then it was said, "*The treasures of Tut-ank-Amen's Tomb were no more precious to the Egyptologist than are these literary treasures to the world of scholarship.*" It is even said that these writings could be traced to the Greeks, Syrians, Egyptians, Abyssinians, Hebrews and many of the other writers and recorders of history in ancient times. As one writer was quoted, "*As a lawyer might say who examines so much apparently unrelated evidence . . . there must be something back of it*" (*ref. 31*).

This shows that the Bible was given to those whom God chose to inspire to write it. For the most part, these men couldn't even write, and that's where the scribes came into play. These men were given the Word or guidance through God's inspiration. Many of them were simple fishermen or corrupt tax collectors and politicians, but God still used them. The works of the Scriptures are what God gave us, and then the rest of the storied history is documented through other writings over time. I truly believe that the reason for the attempt to keep us from knowing any of these truths goes back to what I have said several times over; the events of the Bible are too big for most people to grasp, especially those of little faith who claim to have much. That's why there's so much effort to keep knowledge to a minimum in order to subdue the questions of what, where, and when.

FOLLOWING THE BIBLE

Moving further, we now turn to the time period at the end of the lives of the disciples who walked with Christ, and moving into the years of those who accumulated the writings of the Bible and the history of those who evangelized the Word of God in the early years of Christianity. It's just like we still do today. As I mentioned earlier, when we want to know the truth about Abraham Lincoln, we gain the best knowledge from reading the historical accounts of those who lived during the late President's life or to those shortly after his life who began to interview and record the events that took place. The same concept applies to those surrounding Jesus and the men who wrote the Gospels and the historical books that followed shortly thereafter.

During the final years of the apostles, their lives were recorded by many of those men who were disciples and followers. For example, Clement of Rome was known to have been with the disciples during their years of preaching as we read by Peter's reference:

> *And I intreat thee also, true yokefellow, help those women which laboured with me in the gospel, with Clement also, and with other my fellowlabourers, whose names are in the book of life*. I Peter 4:3

Clement went on to write several letters of historical text, and was even one of the early church leaders in early Christianity. Clement, as well, recorded some of the early years of the disciples, along with many letters to some of the churches that had been established. It's also believed through many writings that Clement was there with Paul and Peter when both men were martyred in Rome. Men like Papias, Polycarp, Ignatius, and Prochorus were said to have worked with or were discipled by John. Church historian and early Bible translator Jerome quotes some of Papias' work in his writings of *Lives of Illustrious Men*, which says, *I used to inquire about what Andrew or Peter had said, or Philip or Thomas or James or John or Matthew, or any other of the Lord's disciples, and what Aristion and John the Elder, disciples of the Lord, were saying*. *For books to read are not as useful to me as the living voice sounding out clearly up to the present day in the persons of their authors. (ref. 32)*

Another early church leader and writer is Eusebius, a Bishop of Caesarea in Palestine, with his works of Church History or Ecclesiastical History, which gave us a clearer glimpse into the lives of those who served in the 1st century through the 4th century. He gave us the first of the accumulated writings of the witnesses of those who served in the church and many of their fates. Others came along throughout time who combined many of the early writings and went on to record them into volumes, such as 10th century writer Symeon Metaphrates. Jacobus de Voragine, a 16th century writer who came along and pulled many of these writers together with his incredibly popular volumes of books called *Legenda aurea* or *Legenda sanctorum*, or better known as *The Golden Legend (ref. 15)*. The most well-known of the early collectors of those who served and were martyred for the church was John Foxe, a 16th century writer, and his works known as *Actes and Monuments of these Latter and Perillous Days, Touching Matters of the Church*, or later at *Foxe's Book of Martyrs (ref. 33)*. Again, all of these writings were a collection of many of the early church leaders and their accounts of the events of those men who served in the church in the early years, from the time of the apostles until the days of their contemporaries.

Amazingly, most of these writings can be pieced together with other clues to reveal hidden gems such as the locations of early church disciples. These can be physically verified; such is the case with this discovery. That's where this story takes a more realistic turn. The writings of the men God chose to record His Word gave us the perfection of life, salvation, and living instructions we know as the Bible. Just as we turn to historians of the days of Lincoln and Washington for the truth on their lives and what happened in the time to follow, the same applies here as we turn to those who walked with the apostles, witnessed their work, and then continued on bringing the history of what has been witnessed, and the truth in what God gave us. Again, this is no different than with the Essenes who copied the Scriptures in Qumran, as well as the historical writings that give us the continued stories of the Bible.

Let God be your guide. And remember this: God is much bigger than the small little boxed edition that the world has put Him in. He moved mountains, wiped out civilizations, and even created this world in just six days. Of course, that is if you believe big enough. In most lives, people look upon God as an "only when you need him" god. To me, and those who allow God work in their lives fully, will quickly see that God can do ALL things at any time to complete His need or purpose. So, who is your God? The one that the world through man has created for us, or The One who can do all things . . . even when they seem impossible?

The theological world has put so much time into disproving much that the early church leaders have written, as if they know better. For example, Jerome confirmed that Matthew first gave his Gospel in Hebrew for those who were circumcised and believed in Christ. Later the Gospel of Matthew was translated into Greek, but that author is unknown. Jerome had one of the last opportunities to see the original form of this writing, as given to him by the Nazarenes. He even states there was a preserved copy at that time being stored in the library at Caesarea. In other words, these early church leaders, translators, and historians knew the facts and lived much closer to the times of those who had the original texts, or they were able to collect the writings of many of those early believers.

To finalize this chapter, we took these writings, the ancient manuscripts of the Ethiopian's history, and then visited with eyewitness observation and excavations to complete these discoveries and determinations. I believe following the Lord's lead will take you toward the tasks that He wants to be shown or completed. Even though modern man may claim to know more, I still believe God has the answers, and I will continue to follow His lead to get to my next destination or discovery. What about you?

Chapter 4

Tradition vs. Truth: God Opens the Door

Ancient Ethiopian painting of the Apostle Matthew

THE ANCIENT WRITINGS THAT record the whereabouts of the apostles over the centuries can give us incredible clues to open up doors when locating a physical place and artifacts of the particular search. Looking into these writings, in order to locate where Matthew preached and was martyred, we first had to establish a general location. With that said, the writers all state we needed to first look into Ethiopia. Looking where the theological world has slapped the nameplate of India, Iran, Turkey, or other locations for Ethiopia simply sends us into a worthless confusion. This method can lead to many dead ends. I truly believe this is the same concept with Noah's Ark. We search where man has decided it must be, with nothing to show for over two thousand years. In our efforts in this expedition, I decided to take what both the Bible and the ancient writings all say when it came to Ethiopia. Let's go where we know Ethiopia truly was and is. It's amazing what will happen when you follow the writings of old, and not try to re-write what has already been written. That's a fact, which you will see in this exploration and discovery.

Theologians use their own belief to lead us down the road they want instead of the world God has introduced us to. In their world, Ethiopia is Persia or India, until it talks about people of a wilderness land, primitive, or untouched. That's when they will say that we are talking about the same Ethiopia we know today. Listen, this has caused a confusion that is laughed at by the world and has turned possible believers in Christ away from receiving Him as Savior. Maybe that's the goal in the first place, because we know who the author of confusion is according to the Scriptures.

When confusion is thrown into the calculation, we see a disbelief in the Bible and God, which explains why the morals of this world have degraded. This confusion also makes us question other locations, such as Israel and Egypt – are they the same today as they were in the Bible? The answer, of course, is yes, so we need to stop the chaos, confusion, and madness, and accept that Ethiopia is the same place of today, even though it may not be the people you want them to be. In actuality, as I've written in my book *Guardians of the Secrets Book I (ref. 34)*, Ethiopia and Africa is the location God speaks of in His creation of the world, so let's accept the truth and move forward. Science and history prove this fact, so we must simply accept it. Speaking of this, we also need to clarify when we speak of Ethiopia in the ancient days, there were only three countries in the original continent of Africa. Phut (Lybia), Mizaim (Egypt), and Cush (Kush) was Ethiopia, which was the largest part of the continent. It wasn't until later when countries started to form and divide the continent when Ethiopia became smaller. Thus, the continent is the birth of life, and the birth of mankind. I'm not getting much into this within this writing. If you need more, read *Guardians of the Secrets Book One* for substantial insight on this.

This same idea has come to light even more in my efforts in the country of Ethiopia. I have personally witnessed confusion by outside individuals coming into this African nation and continuing the false claims of a curse placed upon the Ethiopians by Noah upon his son Ham in the Bible. This outright lie has caused those claiming this theory to somewhat control those they share it with in Ethiopia. If those individuals would simply read their Bible, they would see a curse placed upon Ham's son

Canaan and not Ham who settled the land of Ethiopia. Again, this is simply a way for one group of people to control another group of people, which adds to the confusion of God's Word.

TRADITION VS. TRUTH

Tradition versus truth is a topic that can be heated in many ways. For example, one faith claims that another is doused in tradition, while they follow tradition of the same number of songs on Sunday morning, same time allotted for preaching, and the same type of music that the pastor prior to him had instilled within the church. People have to sit in the same pew each week, as if they own it, while others are disgusted when they see drums setup on the altar. Come on people, when did pianos come into the Bible? Drums have been used, as trumpets and cymbals have, since the early days of the Old Testament, and nowhere do you find pianos, organs, or even sound systems. We use them because God has allowed the invention of them to help the church in its worship.

But then you hear a suit and tie is required to walk into the church, or a dress must be worn by the women, again, tradition has been set. In reality, men should be in robes, as well as the women, and heads should be covered when entering the church and going into prayer. Now that would be real tradition to the Bible text. It reminds me of the church I first pastored. God led us to build our sanctuary to resemble Capernaum, and it was beautiful. Those men and women who worked on it were truly blessed in their efforts. But, within a year, a long-time, so-called Christian walked in, looked around, and proclaims, "This doesn't look like a church. This is a little over the top wouldn't you say?" In truth, this would be more like the look of the early church that Jesus preached in, but tradition has made us foresee a church as a white building with white walls inside, filled with pews or chairs, and a pre-set time for the music to be played and the pastor to preach before people grow weary of the Word and need to get to lunch before the crowd gets too big.

Do I have it right? That's the way the church has been and will be, and the only thing changing is the amount of time the Word of God is preached is being reduced to a mere matter of minutes, as other activities have cut it down to a quick, prepared reading and simplistic sharing to connect us all to feeling good for the day. That is now the trend, and this will continue to progress over the time to come as we turn to a tradition of appeasing the needs of the people rather than sharing what the Holy Spirit lays on the pastor's heart, and God takes over the service into a period of repentance and salvation in Christ. The message heard on Sunday mornings used to be convicting and was talked about all week. In the world today, most people couldn't tell you what the service was about by the time they eat lunch.

So, what does tradition have to do with truth, which is the heading to this chapter? Tradition has been accumulated within the church for centuries, and this has been dubbed by the outside world as legend. In reality, it can be one and the same. In the orthodox church, there are traditional stories that have been added to the biblical text. Even though many of the stories we hear may be true, they are tradition that has been added over time. For example, one story in the orthodox faith is that of Saint George, who slew the dragon (dinosaur) with his spear. George was historically known to be a man born in Greece but served as an officer in the Roman army. It was estimated that he lived in the 4th century and was martyred by Emperor Diocletian when he refused to renounce his Christian faith. Although this story of Saint George may be truth (and most likely is) it was a historical event rather than a biblical event. But in many orthodox beliefs, it has been written into their Bibles when it should have been a historical text. Religions around the world have tradition written into their writings, but in the Christian biblical text there is a laid-out Word that God spoken, or breathed, into a small group of people to record for us the history, the prophecy, the Salvation, the work to be completed, and the lifestyle of those who seek eternal life in heaven. Anything outside of that should be understood as history. It may be the rest of the story, or it may be later history, but it is not of the biblical text.

When it comes directly to God's Word, or God's Truth, we have to understand that it contains a direct divine revelation as canonized text. For example, as my mentor, Dr. Verlis Collins used to say when it came to the commandments of God and Jesus in the Scriptures, he would teach us in Bible college that those commandments should be looked at as law. Law and commandment were one and the same, but in the world, it can be very different. Tradition has a two-fold meaning as well. Tradition of God's Word can refer back to the meaning of this word in the Greek context, "a handed-down instruction." Yes, God's Word has been handed down to us in a traditional way, but that meaning is not the same as a man adding a later event to biblical text and then calling it tradition in the church. Many claim the writings of people such as Clement, Voragine, Foxe, Eusebius, and others are nothing more than tradition. This is not necessarily the truth, because these men were continuing a handed-down work from previous writers to continue the story for generations to come. Men such as Clement and Papias were writing from first-hand accounts they witnessed during their time with the apostles.

SO, WHAT DOES IT ALL MEAN

As we bring this chapter to an end, we must realize that sometimes tradition and truth can be the same. But the best description would be the tradition of passing on God's history of the church, then written down as truth over time for our generation to witness in a life-changing way. On the other hand, tradition of post-biblical events is not to be confused or included in the biblical text, but rather should be written apart from God's Word to separate biblical truth from historical truth.

This has been a topic of division for many centuries. It has also been a topic of confusion to the world. God's Word was given to us ultimately for the redemption of mankind by God's grace through faith. Confusing this with additional stories and mixing them into the canonized text only makes that faith become more watered down and misunderstood by the common reader. We need to stick to the facts; there is God's Word, and then there are other texts that will support the Bible if they are parallel to its teaching that will continue the story following the biblical ending to the text.

When it comes down to the nuts and bolts of the truth, simply know the Bible is perfect and is the inspired Word of God. Anything else is simply history or ancient tradition, and discernment is the best way to distinguish the two.

Chapter 5

True Calling Begins

MATTHEW BEGAN AS A tax-collector and turned into a disciple of the Savior of the world, Jesus. He was hated because of his work with the Romans, and now he was loved by the only One that really mattered. When Matthew left his world behind, he ultimately was called to do much more than he would have ever imagined. From a tax-collector who took from and cheated his own people to a man who would spread the Good News of Jesus to world is a big step for not only Matthew, but anyone.

Think about it, he left his wealth, his family, his world behind. He then stepped into a life of only the possessions he carried with him; he ate only bread and fish or whatever was available. He walked from place to place and was either surrounded by those who would have been curious to hear what Jesus was saying, or they were ridiculed, mocked, and run out of town. Not the life that you and I would think about going into.

As time passed, Matthew and the other disciples were taught the Word of God, filled with the need to receive Christ into the lives, and share Jesus with the world. To all they encountered, they would have been teaching the fulfillment of what the Savior would do for them if they followed Him. It's at this time, when all was well, that Judas sold out to the Pharisees, Jesus was beaten and mocked, and was crucified at Calvary. This is when the life of Matthew and the others truly changed.

THE ARRIVAL OF THE HOLY SPIRIT

Confusion may have been the best word to describe how the Bible depicts the state of those who surrounded Jesus at the time of His death. Many of the disciples were questioning their decision, while others were trying to figure out what was going to happen next as they had returned to Jerusalem. It was then that God stepped in with the next matter that needed to be fulfilled. As the believers gathered in an upper room (about 120 according to Acts 1:15) Peter began the process of explaining the need to fill the place of Judas. Matthias was the one chosen, and the twelve were fulfilled as the real work was about to begin.

The table was now set for the sending of the Holy Spirit as they were all gathered together. Acts chapter two gives a clear and precise description of the events of the day of Pentecost, and one can only imagine the sight that was happening with all of this taking place amongst these followers of Jesus. Finally, the day had come to the fulfillment of Christ's teaching to those believers, as the third of the Trinity had arrived:

> *And when the day of Pentecost was fully come, they were all with one accord in one place. And suddenly there came a sound from heaven as of a rushing mighty wind, and it filled all the house where they were sitting. And there appeared unto them cloven tongues like as of fire, and it sat upon each of them. And they were all filled with the Holy Ghost…* (Acts 2:1-4a)

It was now all in place. The promise of the Holy Spirit of God had filled those believers, and it was now time for them to go and fulfill the commands of Jesus:

> *Go ye into all the world, and preach the gospel to every creature.* (Mark 16:15)

> *Go ye therefore, and teach all nations, baptizing them in the name of the Father, and of the Son, and of the Holy Ghost: Teaching them to observe all things whatsoever I have commanded you:*

> *and, lo, I am with you alway, even unto the end of the world. Amen.* (Matthew 28:19-20)

> *...ye shall be witnesses unto me both in Jerusalem, and in all Judaea, and in Samaria, and unto the uttermost part of the earth.* (Acts 2:8)

This would have been both an exciting and a scary time for the apostles. They all knew the mixed reactions of Jesus' teaching, and they all knew the risks that were in front of them as they departed out into the world to teach and preach the message of Jesus. This is where the real story of Matthew begins. This is where each disciple forgets their personal past and takes the teaching of the Messiah into the world, just as He had commanded.

There were many accounts of Matthew and his early teachings, while later writers may dispute some of those closer to the days of the apostle. Irenaeus (*Against Heresies*) wrote that Matthew began his ministry going and preaching amongst the Jews in Judea before venturing further out into the world. This is confirmed again by Clement of Alexandria, which makes perfect sense considering this was new for all of them. They were all most likely preaching and sharing within their surrounding areas in order to perfect the message and the consistency of what was being shared by each of them.

From this point on, many of the disciples began to spread out into the world as commanded by Christ. Many ancient manuscripts record the history of many of the disciples after they followed the command to go out into all the world. Matthew was said to have gone from Judea into Parthia (near modern day Iran) and preached throughout that region. He was then said to have been in Media, Syria, and Persia before going into Egypt and then into Ethiopia. You read that right - Ethiopia was where Matthew was said to have finished his ministry of sharing the Gospel.

THIS JOURNEY BEGINS TO TAKE SHAPE: THE ARK OF THE COVENANT

(L to R) Bante, high priest of Gondar, and Dr. Rankin study ancient manuscripts

In 2013, while I was working in Ethiopia on a completely separate research project, I decided to return to a very high-ranking priest, Aba Like-Likawint-Ezera-Hadis-Yemetsahft-Gubena-Memhir, whom I had become acquainted with in friendship earlier in the year. Many believed I would not get an audience with this man the first time; even my great friend Misgana, or my guide, and fellow researcher, Bante thought this to be impossible because of the importance of this wonderful man. As I returned to the Aba's compound in Gondar with Bante, I was meeting in his private quarters with him, and among other things, asked him about the truth of Matthew ending his life of preaching in Ethiopia. Again, it was a piece of preaching that I had heard my pastor speak of fifteen years earlier, and it seemed to stick with me about Matthew coming to Ethiopia, as I continued a little more into the research in Bible seminary. After a stare for a few seconds, his reply was, "You are looking for a small village in the north near where the old Axum kings would sit on their thrones." I was very familiar with Axum, Ethiopia. This is where the foundations of early ancient Ethiopian kingdoms were based, and the center of the noted Church of Mary of Zion where the Ark of the Covenant is housed and guarded. I had already done a great deal of work and research there, but it never crossed my mind to begin looking there. I'm not sure why I hadn't thought of that before.

He pulled out an ancient manuscript from a collection he had from his hidden books that I had seen him with that were stored away in a secret location in his study area during a previous encounter we had. He started by saying to me, "You already have a very important part of this truth you have written about, but did not understand that it was a part of what you are asked me about today." He was referring to my book, *Expedition Ark of the Covenant* (formerly *Jesus in Ethiopia*).

He went on to say, "You will also find writings at Tana Kirkos, and in Axum. Keep searching, you will find what you seek. As I spoke to you before, follow where God leads you, not where man wants you to go."

Let's step back for a brief look at this discovery before we move on, because it really does play a huge role in this important journey. Many years earlier, during my first expedition into Ethiopia, we were a part of a team simply on the trail of evidence of the Ark of the Covenant coming to Ethiopia. After seeing the evidence about the Ark's whereabouts with King Josiah's cry-out to the Levitical Tribe to bring the Ark back to Israel in 2 Chronicles 35, we also see in the same Scripture an Egyptian Pharaoh marching his troops to go into battle with a completely different group of people. First, let's look at the Josiah's plea to the Levites.

> *And said unto the Levites that taught all Israel, which were holy unto the LORD, Put the holy ark in the house which Solomon the son of David king of Israel did build; it shall not be a burden upon your shoulders: serve now the LORD your God, and his people Israel,* (2 Chronicles 35:3)

From here, Josiah later confronts the Pharaoh, and the Egyptian king sends messengers back to Josiah to tell him to keep his distance, and that he has no business with him. It was at this point we see the Pharaoh make an important claim, and that was that he was getting his commands from God. Yes,

Jehovah God, and how do we know that? Because, not just once, but twice does he proclaim this statement in the same Scripture.

> *I come not against thee this day, but against the house wherewith I have war: for God commanded me to make haste: forbear thee from meddling with God, who is with me, that he destroy thee not.* (2 Chronicles 35:21)

We also know that the only way that Pharaoh Necho was getting his commands directly from God was through the Levitical priest's visit with God when He would descend upon the throne (mercy seat) atop the Ark of the Covenant. This statement would be the last primary mention of the whereabouts of the physical Ark. After this comes another incredible series of historical and scriptural clues.

The evidence would lead, at this point, to the Ark physically being held somewhere in Egypt. During the same time period, there is great evidence of a temple that once existed on an island in the Aswan Valley of Egypt, on the Nile River, called Elephantine Island. There have been remains excavated of a Jewish Temple with measurements the same as the temple of Solomon. There is even a room in the far backend of the structure that measures the dimensions of the Holy of Holies where the Ark would have been housed. It's estimated that around 410 B.C., this temple was destroyed, and the Levitical tribe simply left it behind, disappearing, along with the Ark. It was at this time that monks on the forbidden island of Tana Kirkos on Lake Tana (Ethiopia) claim that the Jewish tribe brought the Ark there in a makeshift tabernacle. The Ark rested on the high, guarded cliffs of this island for hundreds of years. Then in approximately 315 A.D., one of the well-known powerful kings of Ethiopia, King Ezana, became a Christian and proclaimed this faith back to his kingdom. It was also around this time that the king and his soldiers traveled to Tana Kirkos Island and claimed the Ark of the Covenant, then transported it and the two silver trumpets of Moses (mentioned in Numbers chapter 10) to his kingdom in Axum, where it still rests in waiting today.

Many claim the Ark could not have left Israel, but we know from reading the Scriptures in 2 Chronicles (mentioned in this chapter), and from the prophets, that this is simply not true. Isaiah states:

> *Woe to the land shadowing with wings, which is beyond the rivers of Ethiopia: That sendeth ambassadors by the sea, even in vessels of bulrushes upon the waters, saying, Go, ye swift messengers, to a nation scattered and peeled, to a people terrible from their beginning hitherto; a nation meted out and trodden down, whose land the rivers have spoiled! All ye inhabitants of the world, and dwellers on the earth, see ye, when he lifteth up an ensign on the mountains; and when he bloweth a trumpet, hear ye...In that time shall the present be brought unto the*

> *LORD of hosts of a people scattered and peeled, and from a people terrible from their beginning hitherto; a nation meted out and trodden under foot, whose land the rivers have spoiled, to the place of the name of the LORD of hosts, the mount Zion.* (Isaiah 18:1-3, 7)

This directly refers to Ethiopia, and the description of Isaiah of this country is perfect compared to what we know today. The skies swarm with eagles, and the rivers run through this country of Ethiopia, including the biblical river Gihon (Blue Nile). It goes on the speak of the *vessels of bulrushes upon the waters*, and Ethiopia, especially Lake Tana, is covered with papyrus, or bulrush boats every morning with fishermen.

Ethiopian fisherman in a Tankwa papyrus (bulrush) boat on Lake Tana

Isaiah speaks of tall and smooth skinned people; whose land is desert-like and the waters of the rivers are gone and what waters exist are not good. He speaks of the mountains of Ethiopia, and then he tells of the blowing of the trumpets - not shofars, but the trumpets that have always been blown when the Ark was moved from place to place, which are housed in the treasury next to the church of the Ark in Axum today. Then Isaiah speaks incredible telling words beginning in verse 7 when he begins with, *In that time shall the present be brought unto the Lord of hosts*. This phrase is speaking directly of a very important gift, or exalted gift as translation offers, to Jesus (*Lord of hosts*) where He will take up His throne, or His dwelling place as we see at the end of the verse (*to the place of the name of the LORD of hosts, the mount Zion*). The only item ever in the presence of the throne room, or God's dwelling place, was the Ark of the Covenant with the throne of God atop.

This reference is again made by Prophet Zephaniah:

> *From beyond the rivers of Ethiopia my suppliants, even the daughter of my dispersed, shall bring mine offering*. (Zephaniah 3:10)

The prophet shares the same idea, mentioning the rivers of Ethiopia once again, speaking of the dispersed, or the Jews; Ethiopia is filled with very distinct groupings of Jews in the mountains of northern regions still today. Then he shares that they will bring: The offering. This, again, is the exalted gift as Isaiah spoke of. This brings us to 2011 when we were on an expedition to Lake Tana to search out the evidence of the Ark of the Covenant on Tana Kirkos Island, a somewhat forbidden island of the past located on this massive expanse of water.

While on the island, we trekked through the jungle, which only the men were allowed to enter, and then came to a certain area where we were asked to take off our shoes because we were about to walk on holy ground, as it was described to us. After walking another 50 yards, the high priest on the island began to tell of the rock platform in front of us was where the Ark rested before King Ezana came and took it to Axum. At the time we were there, the platform was covered with grass, leaves, and dirt.

Another man with us, Bob Cornuke had gained access to the island some fifteen years earlier, and began to show us the holy of holies pole holes that he had discovered, matching the dimensions of the biblical place where God's throne was kept in the tabernacle *(ref. 35)*. On the way to this spot I had begun to feel quite ill, and eventually collapsed to my knees. It was here that I simply began to pull the grass and throw the dirt and leaves off of the side of the cliff. After I completed doing this, which was much to the dismay of the high priest at first, I discovered seven indentions, or footholds in the rock.

Holy of Holies pole hole on Tana Kirkos Island

Ark of the Covenant footholds discovered by Dr. Rankin in 2011

I pulled a small tape measure I had been carrying in my backpack and began measuring them with the help of a friend of mine on the journey, Dave Kochis. Right where they had always described the Ark as resting, the dimensions of the rectangular box measured the dimensions of the Ark of the Covenant of the Bible, right on the exact place they had always professed it was located. They had simply never fully cleaned the area to find out.

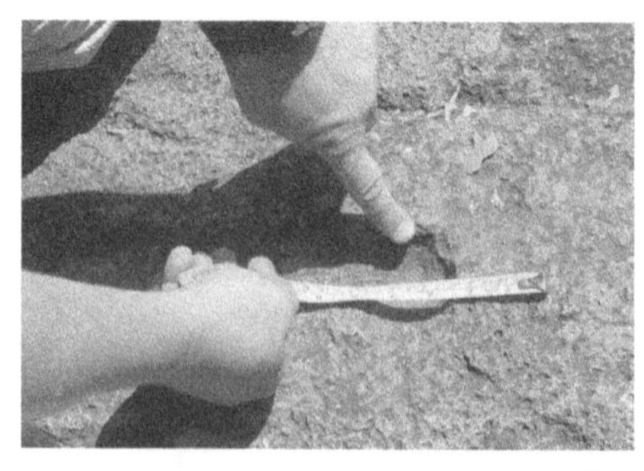

Measuring the footholds of the Ark of the Covenant

It was also in the center of where the pole holes were located that Bob had found years earlier. To our knowledge, this is the only evidence of the Ark's existence besides what we have in the Bible as proof. The high priest, after he saw what I had found, got down on his knees to touch the indentions in the stone platform.

Sacrifice blood bowl, Ephod harness, meat hook for draining blood into the bowl of sacrifice, meat fork for burnt offerings with almond bud symbol at top

We were also taken back to the ancient stone treasury located about fifty yards from the site of the footholds of the Ark of the Covenant, where they were housing over four hundred ancient manuscripts mentioned in the Bible, biblical texts, and some of the historical texts discussing the apostles and others. They also house the blood bowl for the dipping and then sprinkling of the blood on the throne atop the Ark, as well as the harness that held the ephod of the high priest. Within the deteriorating structure are also the hooks for the meat to drip the blood into the bowl, the large forks for the meat for the burnt offering sacrifice, as well as the laver, or bowl, for cleansing the high priest. In 2017, the metal items were scientifically dated, and they are actually from the time period of the Ark's existence on Tana Kirkos Island, and beyond that period.

As we push forward two years to my meeting with the high-ranking priest in Gondar in 2013, his answer to me was very direct. I asked him what the name of the village I should be looking for, and he quietly responded, "Do your research and you will find it. You have found much; this will be right in front of you." He pulled out another very old manuscript, and began to show me some clues of things that must be in the place that I looking for in order for it to be the right place. We finished talking and eating chick-peas together, and Bante and I stood to shake the priest's hand. He stood, and then nodded his head to me, as if to confirm our friendship and the path I was embarking on, just as he had done before. Bante walked away with a smile on his face; he turned and looked back at me as we trekked down the pathway, and said, "Mr. Jim, I cannot believe you were able to meet with this man the first time, now you are friends. He trusts you, I can tell. He freely gives you information, but not all the information. He wants you to go forward to search for truth. Even when I see him today here in Gondar he calls me by name because of you. It's amazing!" You could tell he was proud to be a part of the work and the research, and I was happy to have him a part of it as well. But for now, the trek goes forward to fill in the blanks. I believe that's why the priest won't give me all the information; he feels that when I see the evidence come together, then it will make the truth become more believable.

Chapter 6

Acts: Philip and the Ethiopian

AT THIS POINT, YOU have to be wondering what all of this has to do with the discovery which lays beyond in these pages!

The Ark of the Covenant, the wise men, old historical manuscripts, Philip, the Ethiopian Eunuch - what does it all mean? Understanding that the Ark of the Covenant would have been considered to be in the treasures of the Ethiopians at this time, we now start to see a clearer picture to why so much was written about the eunuch in this Bible event. Stay with me, because it all comes together in a spectacular way as this journey progresses.

ETHIOPIAN EUNUCH

The Ethiopian eunuch is one of the most discussed passages when it comes to the salvation and baptism of an individual in churches. But this account with Philip the Deacon and the Ethiopian man offers us an incredible story that is most of the time swept under the rug. This account of the Ethiopian eunuch is another wonderful example of a biblical event presented in great detail, but with the true explanation most often overlooked. When Luke described the events of Philip and his ministry in Acts 8, he gave a tremendous amount of time to one particular person in this passage. As I mentioned above, the Ethiopian eunuch has been the example for many pastors over the ages as the perfect standard for salvation and the immersed water baptism. What could possibly be overlooked is Luke's apparent clue to the importance of this man and the story's relevance to the Ark of the Covenant's location, and eventually Matthew?

Whenever I come across a biblical event that gives great detail, I become inquisitive why God allowed this much information to be given. Most Scripture referring to various quick mentions of figures are very quickly shared and then it moves on. This particular passage in the Bible contains great content and detail to who this man really was. I had always pondered this detail, but had fallen into the accepted tradition of what this Scripture was intended to be used for. That tradition, of course, was only to explain salvation and baptism to a possible new believer in Christ.

And the angel of the Lord spake unto Philip, saying, Arise, and go toward the south unto the way that goeth down from Jerusalem unto Gaza, which is desert. And he arose and went: and, behold, a man of Ethiopia, an eunuch of great authority under Candace queen of the Ethiopians, who had the charge of all her treasure, and had come to Jerusalem for to worship, Was returning, and sitting in his chariot read Esaias the prophet. Then the Spirit said unto Philip, Go near, and join thyself to this chariot. And Philip ran thither to him, and heard him read the prophet Esaias, and said, Understandest thou what thou readest? And he said, How can I, except some man should guide me? And he desired Philip that he would come up and sit with him. The place of the scripture which he read was this, He was led as a sheep to the slaughter; and like a lamb dumb before his shearer, so opened he not his mouth: In his humiliation his judgment was taken away: and who shall declare his generation? for his life is taken from the earth. And the eunuch answered Philip, and said, I pray thee, of whom speaketh the prophet this? of himself, or of some other man? Then Philip opened his mouth, and began at the same scripture, and preached unto him Jesus. And as they went on their way, they came unto a certain water: and the eunuch said, See, here is water;

> *what doth hinder me to be baptized? And Philip said, If thou believest with all thine heart, thou mayest. And he answered and said, I believe that Jesus Christ is the Son of God. And he commanded the chariot to stand still: and they went down both into the water, both Philip and the eunuch; and he baptized him. And when they were come up out of the water, the Spirit of the Lord caught away Philip, that the eunuch saw him no more: and he went on his way rejoicing*. (Acts 8:26-39)

Without any doubt, this is a wonderful historical account of the acceptance of Christ, followed by baptism. But what if there is more to this story? What if this detail was to give us another clue into where the Ark ended up, and eventually to an encounter that this meeting was setting up in the years to come? A key figure in this Scripture was the Ethiopian Queen described as Candace. One of the founding historians of the church, Eusebius, who is responsible for writing one of the earliest records of church history that has survived the ages, wrote "*Ethiopia even to the present day is ruled, according to ancestral custom, by a woman*" *(ref. 10, 11)*. This gives us a look into the evidence in the early Ethiopian church showing the country having women as early rulers. The Queen of Sheba (Saba), Makita, is another example of this.

There's not much known about Candace other than what we can learn from ancient Ethiopian manuscripts which gives us more information to the actual names, away from those given into more common European names used during various translations. After spending time in research into these manuscripts I found that Candace's actual historical name was Queen Mindetke. We also learn within the biblical Scriptures that she had a particular Ethiopian eunuch as her keeper of the treasury, which gave him great authority in her kingdom. Going back to the ancient Ethiopian manuscripts I also found that the eunuch was named Bakos, which we can assume was not recorded when the biblical account was written, or it was lost in time during the translations of the Scriptures.

As Philip arrived on the desert road known as Gaza, we see this Ethiopian sitting in his chariot, confused about what he was reading in a large scroll from Isaiah describing the characteristics of Christ in chapter 53. Was the confusion of the eunuch based on his visit to Jerusalem, only to find it somewhat depressing? He probably had arrived in Jerusalem expecting to see some kind of celebration for the risen Savior. This may have led to his confusion when Philip saw him pondering the words of Isaiah. Without a doubt, this Ethiopian, along with his queen, would have read Isaiah 52 giving a look into the events leading up to, and possibly following Isaiah 53. But then a quick look at a verse in Isaiah 52, may give us a little more insight into the purpose of this eunuch:

> *Depart ye, depart ye, go ye out from thence, touch no unclean thing; go ye out of the midst of her; be ye clean, that bear the vessels of the LORD.* (Isaiah 52:11)

Could this be what the eunuch took as a sign that he was to go to Jerusalem to see if the holy artifact that Queen Candace (Mindetke) held in her treasury, needed to be returned to the Christ at Zion? Scripture tells us that they are to depart and touch no unclean thing, which perfectly describes why Candace would have sent the eunuch, simply because of his purity. And since this eunuch was the keeper of the treasures of the queen, could the reference to *the one who bears the vessels of the LORD* be a direct tie to Ethiopia's claim that they are holding the Ark of the Covenant of the Lord and the Mercy Seat as referenced in Isaiah 18 and Zephaniah 3, as we spoke of earlier?

Let's also take a moment to look closer at the Ethiopian mentioned. Knowing that he was the keeper of the treasury, we can assume that he was no "normal" eunuch. This man had to be much more than later translations of the Bible have made him. During my study in the library at the Hebrew Union College in Cincinnati one particular day, I was addressed by an astute older Jewish man from Israel, a rabbi, who was visiting the college, that saw my close study. I had a stack of books on the table and he inquired about my research. I had the books opened, searching for anything I could find about Old Testament uses of a eunuch in the ancient world; the man asked if my study had anything to do with the Ethiopian in my Bible in the book of Acts Chapter 8. I told him that it did, but I wanted to see what the uses were for eunuchs leading up to the time of the New Testament. He, being a Jewish man of authority himself, as noted by his black gown and kippah he wore upon his head, was quick to point out to me that I may be looking at this in the wrong text. He showed me in another book (of the many he had with him) that the eunuch may be a more important member of the queen's inner realm than what we interpret in the Bible. And by the way, remember this is a Jewish rabbi speaking to me about the New Testament, which I thought to be unbelievable. He said, "You must look at this. He was keeping her treasury. Do you think he was just some eunuch watching over the most important holdings of the country?" I shook my head from side-to-side. He went on, "This man was a man of importance and in the old Hebrew he would have be more known closer to that of a general who would have been castrated for the queen's protection. That would lead us to believe that when this was translated that the Ethiopian man lost a little clout by just being called a eunuch." He went on to say, "This man was most likely an Ethiopian Priest. Maybe even a Jewish man." With the knowledge of what was in Ethiopia at the time and the closer look into what Luke wrote about this man in the Scriptures, it all made a great deal of sense, and that explanation would be more believable regardless the reason this man would have been such a key piece to all of this story, as well to the Ethiopian queen.

Upon a return journey into Ethiopia a few months later, I spoke with Sisay, the young man with whom I built a great, trusted relationship with, to look into this deeper. He came back with the same information, that in the Ethiopian Amharic translation he was known as the "jandereba," or to say, the general or priest. This all makes more sense to the understanding that the queen would never send just anyone to investigate such an important event, but to send a general or priest, who was the keeper or watchman of her treasury and the Ark of the Covenant. Keep this in mind; the Ethiopian man goes to Jerusalem carrying a large scroll of Isaiah. He's searching for the Messiah and to confirm the possibility that now was the time the Ark of the Covenant, or more importantly, the Throne of God (Mercy Seat) is to be returned. Instead, he's confused and begins his journey back to Ethiopia. That's when God sent Philip for one of the most noted conversions in the Bible.

We know, as noted in the Scripture, the Ethiopian accepts the salvation through Christ, and Philip takes him down into the water to be baptized. Once realizing the Mercy Seat was not needed at this time, and with his new found assurance, he returned to Ethiopia rejoicing. He may have been like many Christians today. He may have known Jesus, but through the turmoil he was experiencing in Jerusalem, such as the turmoil in our lives, he felt that somehow, he was lost. As for Philip, the Bible tells us he was caught away by the Lord and ended up in Azotus, many other cities, and ended up in Caesarea, preaching all along the way.

All of this again raises a question why the detail of this story is in Acts chapter eight if there is not more meaning to it. If it were only needed for a salvation and baptism experience, much of this Scripture would not have been needed. Maybe the Lord had more to share than we have been taught to believe. Maybe this gives us a closer glimpse into the importance of this ancient land of Ethiopia and Africa once again. It also opens up the doors to see the importance of Queen Candace (Mindetke) and her role in previous events of the Bible, which we will see more of in the in the next chapter. For now, we see that when the Bible offers us great detail, that may be God's way of telling us to seek Him more.

Chapter 7

The Encounters

AS MATTHEW MADE HIS way through Egypt, as many of the old writings suggests, he preached for a period in arid country, and eventually continued southward, most likely on the Nile River, towards the African nation of Cush (Ethiopia). Since there is no record of how he made his journey, he would have either traveled the Nile, or he would have taken a boat on the Red Sea to the eastern shores of Ethiopia. There were a handful of early church historians who have placed Matthew and Philip (the deacon from Acts 8) together, preaching in the same areas in the Middle East. With that said, you would assume they would have encountered and fellowshipped together in these

times. This would hold true with the next encounter that will happen in this incredible journey.

Before we head into this any further, let me help you understand where much of this content comes from because it's very important to know. I must start by saying that over the centuries, the first written history of biblical events (and even the history of the church) and their writings have been altered by so-called scholarly men who have made decisions concerning what is important and what is not, veering away from those who knew the writers of the biblical books personally or were within a few years of the events taking place. What we have found over the last few years of intense research is that if we go back to the books that were studied by our forefathers of the church, and then actually take the time to go and find out if these places and things are true, it's astonishing the doors the Bible will open up. We found that true in our discovery of the footholds of the Ark of the Covenant, the ancient manuscript of the early missing years of Jesus, the Garden and the Eden of the Garden, Adam's tombs, Golgotha, Noah, and more. We are to use the writings of the past to help us understand it better. Let's again, as I mentioned earlier, look at what the Bible states:

> *Remember the days of old, consider the years of many generations: ask thy father, and he will shew thee; thy elders, and they will tell thee.* (Deuteronomy 32:7)

And then again in Romans:

> *For whatsoever things were written aforetime were written for our learning, that we through patience and comfort of the scriptures might have hope*. (Romans 15:4)

This tells us that we are to use what is given, written to us and for us in the past, especially those of the early church, in order to understand further. In this case, it was to understand what took place after the Bible finishes giving us what we need. Again, the Bible is perfect, but the stories didn't end with the Bible. The rest of the story was still to come. Other men and women over the ages witnessed and continued to accumulate the history, wrote it down, and collected them together, so we could have an account of where the Bible leaves off. As we read in the Bible many times, since there was no Bible to give reference to at the time, the writers and figures in the Bible mention by name, or quote, many other books as historical fact. Many of these books we have used to locate some of our finds, while others followed up the Bible in later years, but all within the parameter of those who knew the truth.

This fact is true with the men who gave us the history of those who served in the early church and their fate. As I have already mentioned before, men like Clement, Papias, Eusebius, and many others were responsible to witness, collect, and ultimately record their accounts of what they knew. Then, over time, it was accumulated, written, and one day published for all to see. It wasn't until later that man believed himself to be more knowledgeable and began to throw out many of these writings. Their reasoning may have been that they did not want to believe these events took place in a certain location, or even that they didn't want to explain them. Either way, pastors and historians have been quoting and using many of these writings for centuries, but most in their audiences have simply assumed it was from the Bible, when in actuality it was from the accumulation and dedication of other writers who preserved this history.

With all this said, many of the men who wrote about the ministry of those who witnessed the work of Jesus' commands and the fates of those individuals were men like Clement of Rome who worked with Paul and Peter, Papias, who personally witnessed many of those disciples, and then historians who accumulated much of the history, such as Josephus, Eusebius and others. One of those who took the writings of these men and collected them all together was Jacobus de Voragine (1230-1298). He was a church chronicler and the Archbishop of Genoa, and he was responsible for many works over the years, but known more in-depth for *The Legenda Aurea* (or *Legenda Sanctorum*, or *Golden Legend*). This was a massive accumulation of the prominent workers of the church and their fates he collected into one multi-volume work. He included the works of many of those I have mentioned before to finish his own writings.

One of the more well-known of these men who collected the history of those who greatly served in the spreading of the Gospel was John Foxe (1516-1587). Foxe was a church historian and martyrologist, and he was the author of the most quoted of the collections of the fates of those in the church with his writings of *Foxe's Book of Martyrs* (originally titled, *Actes and Monuments of these Latter and Perillous Days, Touching Matters of the Church* or simply *Actes and Monuments*). I have sat through many sermons in many denominations of churches and heard preachers quote numerous things written within this book, which again is an accumulation of men like Clement, Eusebius, and Voragine.

In all, these men gave us a history that we can go back to and research. A history that can be followed, and eventually (and prayerfully) discovered in order to give us a closer reality to the Bible and to those who wrote and shaped the accounts within its pages. Today, too many authorities will claim the impossibility of many of these writings because they have never gone beyond sitting behind a desk or computer in their research. The true amazement in God's lead is when He directs your paths to the facts. If you can take the writings of old, and the Bible, and finally matching them to a location, the results can be completely overwhelming. This can only be true if we can step beyond our limits of what the world decides for us.

THE FIRST ENCOUNTER

Early church founder, writer, and historian, Clement wrote that Matthew traveled to Ethiopia, which really gives a clearer picture of this meeting. Then within the pages of *Foxe's Book of Martyrs* and The *Legenda Sanctorum*, we were able to piece together the full story and finally begin the journey.

After Matthew finished his work in Egypt, he made his way to Ethiopia. It was here that Matthew, because of his previous ministry meeting with Philip the Deacon, sought after a man that Philip knew very well from Ethiopia. Matthew made an acquaintance with the Ethiopian eunuch (Bakos) from Acts chapter 8, in which the apostle was taken to the kingdom of the king and queen. It was here, in a village called Nadabah (later called Mahibre Dego) that adjoined the capital city of the kingdom, that Matthew began to minister. During this time, he immediately encountered two sorcerers, Zaroes and Arphaxat, who were strongly against any work of Matthew in this area, as they would have been threatened by his ministry and the miracles that may be performed. The apostle thwarted the sorcerer's efforts to convince the people to worship them because of their works, and he cast them away from the kingdom. He preached to the people, and through this event led many who witnessed the apostle's encounter with the sorcerers to Christ. The eunuch was astonished how Matthew could simply begin to speak to the people in their own language, and the disciple of Christ told him of the coming of the Holy Spirit within them, they were all given the *knowledge of tongues* to enable them to reach the world *(ref. 15)*.

The apostle went on to another encounter with these same sorcerers who returned, this time bringing dragons, which could have been some type of living dinosaur of the time or even massive crocodiles. Many times, when the word dragon is used, most people believe it to be fantasy by not knowing the history of the word. Since the name "dinosaur" wasn't even a word used to discuss ancient animals until Sir Richard Owen combined the description of a newly found dinosaur fossil in 1841, the former use of such reptiles, or even fossil finds in the previous years, was the description of dragons. So, in this case they brought these reptiles to devour Matthew, but he was able to calm the dragons to sleep at his feet, which infuriated the sorcerers. Again, they were cast out of the kingdom for good by the eunuch (general or high priest), and they retreated far away into Persia. It is written that both Simon and Jude were preaching there, and both of these men *vanquished* the sorcerers for the final time.

DEATH OF THE ROYAL SON AND THE SERMON ON EDEN

After this encounter, Matthew began to preach and share about the former paradise of the Garden of Eden with the people, which I find an incredible piece of history based on our discovery of the Eden of the Garden in my second book, *Guardians of the Secrets Book I (ref. 34)*. With this sermon to the people, it was said that a loud cry broke out amongst the people as word arrived that the son of the king and queen had died. Immediately, the eunuch came to Matthew and brought him to their dead son. As we know from writings in the Bible and other follow-up text, the disciples of Christ went out into the world, as commanded, and were able to perform miracles amongst the people they ministered to. It's also written that angelic beings, sometimes appearing to look like Jesus, made appearances in various forms with the apostles in some cases. In this case, Matthew came to the child, and in prayer, lifted him back to life in the name of Jesus. There were heralds sent out into the people, and into surrounding villages, proclaiming "*Come and see God hiding in the form of a man!*" The people came bringing gold, precious gifts, and demanding that sacrifices should be brought to him. Matthew forbade the recognition and said, "*Men, what do ye? I am not a god, but the servant of Jesus Christ!*" After hearing this, the people of the kingdom took the gifts they had brought and used it to build a great church. It was at this time through this event, all of the kingdom, which was located in the northern parts of Ethiopia, was converted to the faith of Christ, no doubt through the continued preaching of Matthew. It was also at this time that the entire royal family, including the son, Anon, the boy the Apostle raised from the dead, and the boy's sister, Ephigenia, were all baptized *(ref. 15)*.

JESUS APPEARS TO THE APOSTLE

One of the hardest to understand pieces of the puzzle one must overcome is the short-sided thinking of people, when it comes to God. We have to always remember that God can move a mountain if He so desires. Our thinking can't go beyond what we see in the world today, making faith a thing of the past. That's why I feel the world is spiraling out of control with the evil beliefs and lifestyles they live today. With that said, when it comes to hearing stories from many accounts, such as Jesus later appearing to the apostles (or most-likely, sending an angelic being to perform miraculous things in His name) it's truly difficult for the world to wrap their heads around it. It's all in the world's lack of true faith today. I have seen many miracles take place, many times in my own life – sometimes daily, we just need to step back a little to see them. That's why the world has such a problem with believing in the Garden of Eden and Creation, in the parting of the sea with Moses, Noah's Ark, and other events of the Bible. As I've said before, our faith is a god we can control in a little box, not the massive expanse of God that the Bible speaks of.

That's why it's hard for people to believe accounts like what we are speaking of here, including an additional piece in the Ethiopian writings, of an account that took place in the kingdom of Bazen and Candace (Mindetke), where Matthew was preaching. In this writing, during Matthew's time in Nadabah (later called Mahibre Dego), next to the Axumite Kingdom, a beggar was seen making his away around the town, especially near the building in which the apostle was preaching. He was treated badly, as the priests of the church would often have him cast away. There were a series of events that took place as the beggar pursued water and to be fed by someone at the place of worship only to be turned away each time, or mocked by the servants. It was said that a servant gave him a cup of water but broke it intentionally before handing it to the beggar. The beggar took the cup; restoring it and extending it taller and full of water. That's when this account caught my attention. The servants brought him to Matthew who was gathered with other priests near a massive balancing rock. The huge rock was balancing on a very small rock platform, and the previous sorcerers claimed it was the rock of the gods and they worshipped the formation. When the beggar was in front of Matthew, he removed his hood, the apostle recognized the person as an angelic being appearing as Jesus, in which the apostle was said to have fallen on his face in front of Him. The others gathered at the rock were very skeptical at what they were seeing. The writings state that even though Matthew was preaching, and the people claimed their faith in Christ, many still had that old belief in them. Much the same today, as we're not willing to let go of the past to serve God in the present.

The desert area was said to have rocks that bled, but it was actually iron in the rocks and the sorcerers claimed the gods would provide them iron from the bleeding rocks in another attempt to win them over. On this account, the beggar, or the angelic figure appearing as Jesus, took the silver staff Matthew carried with him (made by the people for the apostle from one of the many silver deposits also found in the area) and raised it to the heavens, and to the shock of the priests and deacons, slammed, or plunged the rod into the balancing rock. This caused water to flow from the rock, and into the dry valley below, which brought a lush jungle to the dry landscape. After this event, the beggar, faded away, and Matthew's staff fell to the ground.

LUSTFUL DESIRE TAKES THE THRONE

As the years passed by, we know the king died and the queen's power was taken from her, most likely from being overthrown in battle for the throne which was a common event of that age. We presume this to be true because the new king who took over the throne, named Hirtacus, was not of the faith in Christ, and the events to follow confirm this with the actions of God upon the calling of Matthew. As mentioned, the former king, whose sub-name was Bazen Egippus, and his wife, Candace (Mindetke), had a daughter named Ephigenia. King Hirtacus desired the virgin girl and sought to take her as his wife. Matthew, in a forceful manner, claimed that the former royal family would come and worship in church, and that he should do the same, in which they would discuss the meaning behind a marriage in the sight of God. At this point, the king was filled with joy, but that was to quickly change.

As the people gathered, including the princess, and some two hundred virgins which Matthew had ordered to watch over her, the Apostle began to share the importance of marriage, especially speaking to the princess and the group of virgins. King Hirtacus was overwhelmed with the Apostle's knowledge, and he even praised the preacher of God's Word for what he was giving to the people. But then Matthew turned and commanded silence; as he continued in his message to the people, saying, "*But ye that be here, know ye well that if any servant would take the wife of a king wedded he should not only run to the offence of the king, but above that he should deserve death, and not for to wed her, but for that he in so taking the spouse of his lord should corrupt the marriage joined. And thou the king that knew that Ephigenia is made the spouse of the king perdurable, eternal, and is sacred consecrated with the holy veil, how mayst thou take the wife of a more puissant powerful king and couple her to thee by marriage?*" In other words, Matthew was condemning the custom of taking more than one wife, according to God's definition of marriage. This enraged the king, and he stormed out of the church in a state of *madness*, *insanity*, and *frantic* as Voragine described the scene in his writings. Those that had gathered laid in front of Matthew at the altar as he continued to preach to the people, and more specifically to the princess and the virgins.

THE APOSTLE IS MARTYRED

Martyrdom of Matthew

From this point, it's not specific if the people were still there, but most likely they had all left, because the manuscripts describe the scene as being after the gathering, or after the mass of people had departed. Matthew turned towards the alter and began to pray with his hands lifted up to God. The writings then indicate that King Hirtacus, in his rage, gathered what is called in one translation a swordsman *(ref. 16)*, and then in another more specific period translation called a tormentor *(ref. 17)*. Either way you look at it, this was an assassin for the king. While carrying a sword, he went into the building, stepped up behind Matthew as his hands were raised in prayer, and drove the sword through the apostle's back, killing him. Some writings even indicate that the apostle's body was taken out in a courtyard, nailed to the ground, and beheaded. Most do not share this as fact, but looking at the madness of the king, it may have been entirely possible.

After Matthew's murder, the people of the kingdom tried to overrun the palace of the king, but they were held off by guards, priests, and deacons of the church. It was then that the king attempted to retrieve the princess one more time without success; he then had the living quarters of Ephigenia and all of the virgins surrounded, and set fire to the building with the intent that all inside would die. Instead, Matthew appeared in the fire, and the flames went out from the quarters of the princess, then erupted, and engulfed the palace of the king.

Martyrdom of Matthew with Ephigenia shown at right

KING'S FINAL ACT AND THE RIGHTFUL KING TAKES THE THRONE

As the palace burned, it was written that all of the king's servants and members of his court were killed in the blaze, but the king and his son narrowly escaped. After the fire in the palace, the king's son became hostile and was filled with demons and cried out over the apostle's grave the sins of his father. King Hirtacus, on the other hand, was struck with *foul mesel, leprosy*, as described in Caxton's translation of Voragine's writings, which means an incurable leprosy. Upon understanding that he would be stricken this way, the king took a sword and killed himself, ending his evil reign.

With the death of the king, the people reinstalled the rightful king, which was Anon, the son of King Bazen (Egippus) and Queen Candace (Mindetke). This was the boy that Matthew raised from the dead and baptized, sending a message of faith back into the kingdom. The boy continued to follow the faith in Christ, the faith the apostle had led him to many years earlier. Anon went on to reign as king in Ethiopia for seventy years, and then he turned the throne over to his son.

What makes this incredible account of Matthew's work in Ethiopia even more spectacular is that the Ethiopian people have kept the details of these events a secret for nearly two thousand years. As I have mentioned many times before in my writings, the Ethiopian people are true to keeping the places, meanings, and artifacts hidden away from the outside world. So much so that even those guarding these secrets they hold do not share them with even the village or the people right next door to them. In many cases, they may not know the true facts about what they (and their ancestors) have been guarding for centuries but it has been a given right for them to do so. It's incredible to think that nearly two thousand years later, and even longer in some events, that people have kept hidden these places and things they have guarded by faith until the time that God decides it's time to share them. That's why I truly believe Ethiopians, and other African nations, are the Guardians of the Secrets.

Chapter 8

Piecing the Clues Together

Sisay (left) with Dr. Jim Rankin (right)

THE STORY OF THE whereabouts of Matthew have been written, and it has been told over the centuries, but the real question still persists. Where did this actually take place? Many, over the years, have tried to place these events in countries like India, Saudi Arabia, and others places, which has done nothing but add to the confusion. Theologians will try to divert any attention from the country we know as Ethiopia, when the Bible, in all cases, paints a very clear picture confirming this is the place in Africa, known as the Land of Cush. This is why many locations and items missing from the Bible haven't been found. In order to appease the masses, man has slapped a label on a location and laid claim to it, with little to no factual landmarks or artifacts to confirm the truth. Unfortunately, we simply go with whatever has been said, and bypass any fact about this claim. For me, that's why it's important to find the clues, build the case, and then share it. This is why the priest in Gondar that I mentioned earlier never gives me the whole picture, but only fragments so it will come to life and become reality before my eyes. As he has said to me many times, "Let God direct you, not man." With that said, let's first look at the many clues that we are given in order to locate this place.

When searching for the evidence in this exploration, I had to approach it differently than I have most others. I was working from many angles with this story – the Bible, ancient Ethiopia texts, manuscripts by people like Eusebius, Clement, Voragine, and then searching for the physical locations, while trying to piece together the entire grouping as one. Before, I had a place, but then I had to match the information to that place to confirm it. In this case, I had no actual place, but had to take the information, decipher it, and then set out to locate the place that matched the description, and then ultimately, I could assemble and arrange the puzzle pieces to be put into place. So, with all of that understood, let's get into the evidence that the Lord laid before us.

THE BIBLE'S ENCOUNTER OF PHILIP AND THE ETHIOPIAN

This important clue gives us a look into a timeline of events to set-up Matthew and his work in Ethiopia. Again, as I have mentioned before, it is no coincidence that God allows so much to be written by Luke in this account in the Bible in Acts 8. We learn in this Scripture that God takes Philip away from his work and sends him to a desert road in Gaza. In other words, God sends Philip to the middle of nowhere at the time. Here, Philip is to find an Ethiopian man in a chariot reading Scripture. As a matter of fact, the very important Ethiopian man (who has his own chariot) happens to be reading Isaiah 53, which is describing the coming Messiah and what to look for in Him. Unfortunately, the man is confused because he's just returned from Jerusalem, where he was sent by the queen of Ethiopia, and the city was in confusion and he found no one matching those characteristics. As a matter of fact, he most likely was very confused because of the turmoil in the city, finding out the so-called Messiah had been crucified, but reports of a resurrection had been shared and His return into the heavens had taken place. This was the set up to a great conversion of the Ethiopian to receive Christ and ultimately be baptized. Thus, the man of Africa gets back in his chariot and returns home rejoicing, and Philip is sent by God to Azotus and then to Caesarea.

Here are some things we must first focus on:

- Gaza: If the eunuch was from the Ethiopia we know today, and not from one of the other countries that many theologians will try to place him in, we must determine where Gaza is, which is where Philip was sent to meet the Ethiopian. Gaza, also known as the Gaza Strip, has been a place of political uproar and violence for centuries. It is located west of Jerusalem, and today is a self-governed Palestine territory. It's found along the shores of the Mediterranean Sea on the roadway toward Egypt. This would have been a pathway the Ethiopian man would have been on in order to return to Ethiopia in Africa. Remember, he's in a chariot, most likely traveling with his personal assistants, which means he's not taking a boat on the Red Sea home.
- Man of Great Authority (Bakos the eunuch): He's a man of great authority and not some simple eunuch, as the world has made him out to be. In a way, when the Bible has been translated over time, this man has a contradicting title. Why? Look at the Scripture. He's called *a man of Ethiopia, an eunuch of great authority…*, in Acts 8:27. When the term "eunuch" is placed upon him, it simplifies his rank and lowers him to just a servant. For the Scripture goes on to says of him, w*ho had the charge of all her treasure*, according to Acts 8:27. This man was most likely, as Ethiopian writings share and the Jewish rabbi related to me, a high priest, or possibly a general.
- Candace (Mindetke) the Queen: Here we find the Ethiopian man served under a great queen named

Candace. This is important to the text from later writers depicting the whereabouts of Matthew's final calling to preach.

- The Treasury: This important clue gives us reference to the reasons why Candace sent her high priest, or general, to Jerusalem. Knowing the information about the whereabouts of the Ark of the Covenant and the Scripture backing that Ethiopia has it and will be returning it, is very telling as to why this important man was in Israel in the first place. It's also telling for a certain bond between the king of Ethiopia and Jesus, as well. You'll see more of this later.

PIECING FROM THE WRITTEN MANUSCRIPTS

Now that we have established the Ethiopian eunuch was not going to the far east, but he was heading to the Ethiopia we all know, we can now move further into the clues that we have seen written in chapter 8. During my first journey into Ethiopia in 2011, I had a wonderful encounter with an Ethiopian man in Axum, somewhat a historian, if you will, to the area named Sisay, who has been a partner with me on many of my searches in the past, including walking alongside of me in this journey. In the later part of 2013, Sisay and I sat down to discuss a host of topics, including the Ark of the Covenant in Axum, the Queen of Sheba, and many other facts, legends, reports, and stories. When I brought up the Matthew topic, Sisay's eyes lit up. I mentioned the village, and he felt he knew of a place that would fit the description that I showed him in the writings of the manuscripts of old. He said, "Today, this place I know of, is very small, and they are very closed to outsiders." He went on to say, "At one time it was part of the Axumite Kingdom and was a very large area." Sisay told me that between Axum and this village called Mahibre Dego, there have been many archeological sites linking that this was one very large community.

As Sisay continued to look into this possible location, I continued my research out to other locations that had a history sharing Matthew, or had at one time laid claim to the place the apostle visited. After a few months, Sisay contacted me and said the village elders, and the priest had agreed to let me come into the village. In February of 2014, the opportunity to visit them was granted, and the research was now moving forward into another place that could very well be another write-off on the long list of possibilities of villages of Matthew. Before I went to meet up with Sisay, I visited another far-out community near the city of Mekele, Ethiopia called the Church of Abreha wa Atsebha. My friend Asnake in Lalibela had referred me to this place as a possible site to my search for the Matthew village. It was quite some distance from any kingdom, as the apostle preached in, but it was worth the journey to investigate. Although beautiful, this church's history doesn't date anywhere near the time of Matthew and the events that have been recorded.

Sisay and I met in Axum and began an hour-long drive to Mahibre Dego. As we drove on the rough dirt road with beautiful mountains off to the left, I couldn't help but notice large amounts of people crawling on the ground in the farming fields. I asked

Sisay what was happening, and he responded that the people, after a rain, go to the fields to search for emeralds and sapphires that surface to the top. That is a clue in itself, with the magnificent gold and precious stones that Ethiopia is famous for, this would put this area in the true contention for the actual site. Sisay pointed out to some of the archeological sites that had been excavated on the hillside, as we passed by. We now trekked on, and shortly we drove up a long row of cacti being used as fence line.

Sisay said we had arrived, and we exited the van and started our walk to an old wooden door that entered into the compound and church courtyard. After entering, we turned left, and an old priest was waiting for us, along with a couple of the elders. After a series of greetings, we stepped through another doorway into the courtyard of the church and sat underneath an old fig tree next to the church.

We talked with them for a length of time, both to get information and to gain the trust of the people. At first, they were very standoffish, but later they began to give me more details. At this time, we felt as though we may have taken it as far as we would be able to on a first visit, so we packed up our gear, said our goodbyes, and began walking to the van. The old priest told me that he would welcome me back again upon my next visit. I didn't wait long; I came back in May of 2014 for another visit, and this time they were a little more open right from the beginning. We exchanged greetings, and they shared some writings and artifacts with me from the treasury and in the church. Again, the respect of not going too far was of the essence, so we again left them, knowing we would be back again soon. After a visit with them in November of that same year, again in February of 2015, and a return again in July, I knew we were close to locating something, but there were key pieces still to be determined.

It was February of 2016 when things became very interesting. Sisay and I took a drive to the village, and after meeting with the elders and priests, they allowed us to go wander the area around the compound. The Ethiopians were the originators of tiered farming, and at the top of the rocks, looking down into a great valley, I started to notice that the tiering was much different. As I looked off at the valley, to my surprise was a large lake way off in the distance in this very arid land. I was studying the layout of the tiers, and began to notice the alignment of rocks buried in squared-off patterns, side-by-side, around each of the much larger than normal tiers. Sisay looked at me, both of us simultaneously dropped down to clear these formations. It was quickly evident

Home foundation stones seen clearly in the ground.

that we were looking at ancient foundations to homes. Not just a couple of them, but hundreds of them surrounding the area.

Foundation tiers rimming the hillside

Next to these tiered foundations was a field of layers of rocks, almost resembling a rock pit. Most likely this was where the people tossed the rocks that the homes were made of, clearing the tier as walkways around to various other locations.

Then the true test began - what will we find within these boxed in structures? It didn't take long before we started to find, in every site we dug in, gatherings of

Pottery found throughout the foundations imbedded into the ground.

pottery, and even an ancient coin of the early Axumite kingdom. The men that were there with us claimed there are many coins and pieces of iron found along this pathway where these foundations are located every time it rains.

From here, we moved to another tier, and there were more foundations and more pottery, and we even dug up fragments of iron. This was obviously a large community at one time, but today it's a hidden archeological gem waiting to be researched.

I returned to the village again in July of 2016, and then again in February of 2017. By this time, we had pieced things together, with enough evidence to really make a strong case for this village and what we had discovered within. Each time, we were shown more and more additions to fill in the puzzle, and we were able to locate more in the tiered foundation plots. Then, during a visit in 2018, the priest and one of the deacons sat us under the massive fig tree, and one of them, in the Tigray language, said to Sisay, an uttering of the word, Nadaber. I quickly asked Sisay what they meant by this, and his response to me was to "Wait a minute and I will ask. I am not understanding what he is saying." I was hoping I understood, because this is a name that I know from the ancient manuscripts. They spoke for a few minutes, and then Sisay came back to me. He said, "Mr. Jim, they are saying this is where Matthew lived and preached here at Mahibre Dego, and this church (Enda Tsadkan) was built later to signify the place, and all part of the Axumite Kingdom, and the big city. There is a place over there (pointing eastward), another part of the city called Wereda Adiet, Niadier. This is where the building was were Matthew was stabbed in the back with a sword. Nothing is left there today but foundations, like you see here. This is the place some call Nadaber, or Nadavar. So, this place (Mahibre

Dego) and the place over there (Nadaber) were all parts of King Bazen and Queen Candace's kingdom." Today it is all an area of small mud homes grouped together. So now the entire explanation of the names has been determined, based on the evidence that we had gathered. Their writings within the treasury state the names of these places, and there are a few pieces of artifacts left to that time that hasn't been distributed to other museums or locations over the centuries, but that also is a loose statement, as you will see.

By God's leading, Sisay and I had joined together in Axum, with the help of Bante and the high priest in Gondar, Asnake in Lalibela, and many others, to finally determine the outcome of our find. With all that said, let's look at the evidence to back the ancient manuscripts:

- A Large Community: Voragine stated in *The Legenda Aurea* (*Legenda Sanctorum* or the *Golden Legend*), that Matthew went and preached in the city of Nadabah, which was part of the kingdom of the king and queen, because Matthew's first encounter in the city was with the Ethiopian eunuch that Philip had informed him about while preaching together near Persia. So, first, we are looking for a city or a town located as part of the city of the kingdom, and not a small village. We definitely located that as seen in two side-by-side locations called Mahibre Dego, where Matthew lived and preached (and apparently where the Ethiopian eunuch or high priest lived), and Wereda Adiet, Niadier (Nadaber), where the apostle was martyred. The hundreds of rock foundations where homes once stood is entirely visible today, with incredible collections of pottery, coins, and metal objects to be found within those foundations on tiered platforms throughout the region. It's an archeological goldmine, sitting there untouched for nearly nineteen hundred years.

Massive balancing rock hidden in the village

- Massive Balancing Rock: Two different ancient Ethiopian writings that I was able to see, described the appearance of either Jesus or an angelic figure appearing as Jesus, but coming as a beggar. In both accounts, the books shared the events of the man removing his hood on his gown, revealing Himself to Matthew, then taking the apostle's staff, and striking a massive balancing rock. When He struck it, water began to flow from the rock, pouring into the valley below, filling a portion with water. After nearly three years of coming to the village, the priest took Sisay and I to a large tin building. After unlocking it, they pulled open the doors to let light shine in the building, and what I saw inside left me awestruck. Housed within this structure was a massive rock (seen in the photos) balancing on a small rock below it. When I reached up to touch the rock, it had what I would describe as a slight electrical shock, or more like an adrenalin shock to my hand and my body. It was somewhat overwhelming, as I simply stood there in awe. But as I know from previous discoveries and research, there's much more that needs to be confirmed, in order for this to be more than coincidence.

Water flowing from the foundation of the balancing rock

- Water Running from the Rock: Another key piece to the story of the balancing rock had to be the miraculous flowing of water from it. This entire area sits upon rock, and this balancing rock sits upon a rock on that complete foundation. After standing there staring at the massive rock with my hand upon it, I regained my composure, and the priest motioned with his old wooden staff for me to walk around the back of the rock. As with the first glance of the huge rock, my jaw dropped again at the sight of water running from the back of the rock foundation. As you step toward the front, you could see the remnants of a larger flow of water at one point, but this water was coming out of the rock, then dropping into another gap in the rock. The priest motioned, and Sisay said, "This water now goes back under the rocks, and then comes out over there (pointing to the edge of the rock cliffs)." I shook my head, stepped over the entry to the building housing the rock, and the beginning flow of the water, and stared out over this valley. The only sight of anything not scorched from the sun was the massive lake located at the far end of valley below. The flowing water from the rock was another key piece to this story, but again, there is more to find.

Large lake in the desert valley

- A Lush Valley and Jungle in the Desert: The writings also stated that because of this rock, there had to be a jungle in the desert with a lush valley. As I said, I could stare out over the valley and see a lake in the far end, which for a massive amount of water like this is far from common in this dry region of what used to be the Kingdom of Axum. Yet, there it was, a large amount of water. I turned to Sisay, and I asked him to ask the priest if there were once any jungle-like areas, and was this valley at one time lush and green. First, the old man told him that during the rainy season, the lake is much larger because of the water flowing into it from all areas, and it is very green. It was the next move that shocked me even more. The old priest walked us to the edge of the cliff, and as we looked down, Sisay and I simply looked at each other, and then looked back down. Below us was a lush, green forest of tropical-type trees.

Lush jungle hidden in the valley

I jumped down on the rocks and started a steep descent on the trail into the jungle. It was very lush and dense, and at some points very darkened by the foliage. We were joined by one of our young friends from Axum, Asgedom, and he took my phone and began to film our walk downward into the jungle. I can't begin to describe the feeling of seeing this beauty in such dry conditions, hidden from the human eye, until you walk to edge to look in. As we climbed down, holding onto trees to keep from tumbling toward the end, we

Looking out from the jungle to the fields of crops.

arrived at the bottom. Here, you could see the water from the rock above, flowing into a crystal-clear bathing pool.

Pool of water from the spring flowing from balancing rock above

I stuck my hand into the pool to feel the coolness of the water. Then, jumping from the left was a naked little boy who had heard us coming and was bathing in the pool. Without any doubt, he startled us, as he squatted down near the trees to the left of the pool, and somewhat grunted at us. He moved around in a squatted position and even used his knuckles to balance himself from falling from the rocks, as he circled us to the left, while staring intently. Then, as quickly as he appeared, he leaped into the trees, and ran into the jungle. I asked the man who the priest had sent down with us, if the boy lived here at the village. The man called him somewhat of a "jungle boy" who lives in the area of the trees around us. They don't know who his parents are, only that he has grown up in there and people bring him food.

Jungle boy can be seen bottom left of photo

Wow, another shocking event! But I had to get on track to what we were searching for. This was a perfect description of the jungle-like setting, and through the trees, as you enter into the valley, were beautiful crops of plants, with fruits and vegetables growing everywhere. This piece of evidence had been located, and this one was truly a mind-blowing experience.

Crops growing in the valley next to the hidden jungle

- The Cup Jesus Restored: In the old manuscripts, there was a story of the beggar, who either was Jesus appearing to Matthew, or an angelic figure matching His description, that restored an old cup that one of the servants, in his disgust of the beggar, had broken. The beggar wanted water, but the servant handed him the water and breaking the cup in the process. The cup, according to the writings, was restored by the beggar in front of the entire group of priests and servants. After we returned back up the cliff-side from the jungle below, the old priest had motioned for us to enter into the church with him. As seen in many of Ethiopia's monasteries, there are paintings on the walls leading into the holy of holies in the church. Every Ethiopian Orthodox church has a holy of holies to house a replica of the Ark of the Covenant. This church also had many framed writings, most of them telling of the story of Matthew and others from the region. As they walked me around the corner, there was an enclosed glass case, and within it was a cup, or more like a vase. You can see the original clay cup, with an addition to the top, making it more the size of a vase. According to the writings, Jesus (or the figure appearing as Him), extended the cup in His restoring grip, and filled it with water. This cup has the appearance of a small broken cup, but most definitely with a larger addition to it.

(Left) Extended cup stored within church, (Right) ancient stelae depicting the miracle of the cup

The priest then motioned for me to follow him outside where he pointed to an old ancient stone stelae, simply lying broken on the ground. I walked over to it, cleared away some brush that had accumulated on it, I quickly saw engraved into the stelae a clear representation of the larger cup housed inside of the church next door. By this time, I looked over at Sisay and Asgedom, and said to them that this was amazing evidence that continued to confirm this discovery of the village of Matthew.

- Bleeding Rock: Finally, the simplest of the confirmations found written in ancient manuscripts was the need for iron to be found in the area, visibly seen with rocks which appear to be bleeding containing the mineral. After being taken to the building housing the massive balancing rock, you could see all over the hillside rocks with what appeared to be a dark-orange blood flowing from them. Even the rock that was balancing inside the building contained iron. This hillside appeared to be one large bleeding rock, which is another piece of the puzzle filled in.

THE LAST PIECE OF THE PUZZLE: THE STAFF OF MATTHEW

With all of the evidence building, what could possibly be next? Within all of the writings, we know that Matthew came to the region, preached here for decades, and then was eventually martyred nearby. But what about any remnants of the apostle himself? That was a piece of the puzzle that I wasn't too concerned with, knowing that we are speaking of over nineteen hundred years since those events took place. One thing I have found, when God is leading, the amazement never stops.

During one of my visits to the village in 2015, we sat with the old priests and one of the church's deacons in the compound of the church, as Sisay and Asgedom went back and forth in discussion with them. I would ask a question, the explanation would come, then more discussion would follow. Finally, the old priests stood up, threw his cloak over his shoulder, and walked off towards the treasury. Sisay looked at me, with a surprised expression on his face, and said, "Let's go." As we followed, Sisay relayed to me the priest had agreed to pull a few things of importance out of the small stone treasury for me to see. He brought out some very old ornamental crosses, followed by some of the old books, crowns of some of the ancient priests and kings, and a few other significant items to the church, but insignificant to our research.

It was this point that I turned to walk out of the enclosed area around the treasury, when I saw Sisay's eyes light up. Asgedom said something to Sisay, and he turned and looked at me with a big smile. The priest had been standing back inside the treasury, and seemed to have been deciding on whether to come out or not. Sisay could see him, and had a clear view of what the priest was holding. Sisay stepped away from the door as I stepped a little closer. Sisay was speaking to the old priests in a way to convince him that it's okay if he comes out. He went on to describe me to the old man as a pastor from the United States who has written many good things in the past about Ethiopia. He was trying to comfort the man to bring out what he had firmly grasped in his hand.

It was then that I could see him step toward the door and then out into the area outside the treasury. All I could see was a long, dark gray, metal object protruding from a cloth wrapped and tied around the top down to the middle. Parts of the object were shining, but for the most part it was tarnished. The church deacon and the old priest started to unwrap the cloth around the metal object, and then pulled it off. The old man began to speak, and Sisay became giddy with excitement, while Asgedom's smile grew bigger. Sisay said, "Jim, this is the silver staff of Matthew. Very rarely do they ever bring this out, even for a ceremony."

Dr. Jim Rankin holding staff of Matthew, standing between village high priest (left) and another priest (right)

As I stood there studying the staff, one thing I noticed quickly, was the cross attached to the top of it, as I pointed to it. Without hesitation, the old man knew my concern, and addressed the cross before I even asked. He said, "The cross was added to the staff after 400 A.D., to honor Jesus. This (pointing to the place where the two were joined together) was where the priests of the past put the staff and the cross together as one."

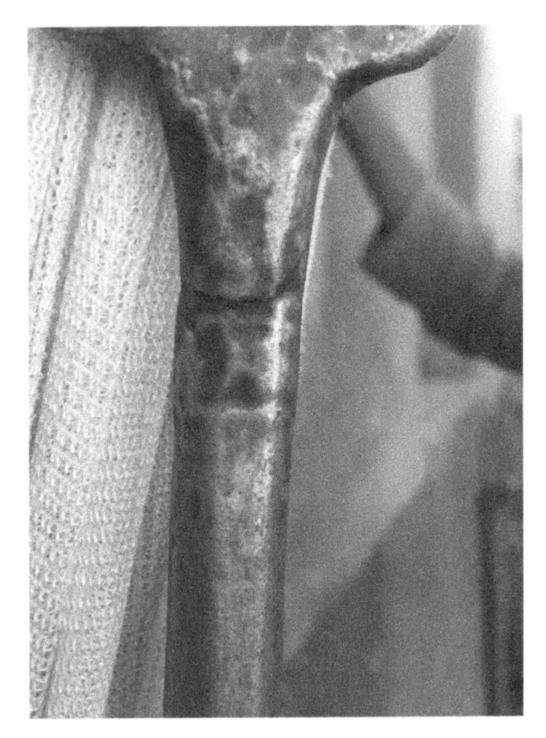

Weld clearly seen connecting the staff of Matthew and the cross added later

I am one of those hands-on people, and I reached to touch it, but the priest drew it back. Sisay looked at me, as I stood there not knowing what to say. This very well could be the staff that was made for Matthew to carry as he ministered and preached to thousands of people, leading them to Christ throughout this kingdom. This very well could be the staff that either Jesus, or a figure appearing as Him, grasped in His hands, revealing Himself to Matthew, and then plunging the object into the massive balancing rock, bringing water to supply the valley and people for centuries. Then, as the priest raised his head to look at me, it seemed as though a peace had come over him. He simply extended the staff towards me, nodded for me to take it, and as I did, he stepped back. It was a moment of trust for him, but a feeling of exhilaration for me. I reached and clutched the staff in my hands, and the feeling was overwhelming.

(Left) Dr. Jim Rankin holding the staff of Matthew, (Right) Village priest, Sisay, and Jim holding Matthew staff

The priest began to speak to Sisay again while I stood there holding and studying the object. Sisay then said, "The priest says, they never let anyone but the high priest hold this staff. No one. He feels you have come in the name of Jesus, and he wants you to hold it, feel it, and know what you are feeling is true, and real."

The emotion that was running through my mind at that moment was overwhelming. I felt a trust had been built over several years of patiently waiting. Not pushing or pressing in a visit or two, but a trust of respect to these men, this place, and for what they hold, and have held here for centuries. My theme verse is Isaiah 40:31, and waiting on the Lord is something you must do when following His lead until the timing is right. That's what I believed happened here as we waited. Sisay had me walk around the treasury enclosure for a series of pictures with the cross, the old priest, and sometimes with himself, as Asgedom took the pictures. Asgedom then wanted photos of him standing with me and the cross. It was a big moment in this investigation, a major discovery in the finding of Matthew's village, and significant trust that has swept over us all.

After a period of time, I handed the staff back to the priest, and he quickly turned to return it to its rightful place within the guarded treasury of the church. Sisay was overwhelmed, as was Asgedom, as Sisay kept saying, "Wow, wow, wow. This was amazing Mr. Jim, amazing. Wow!" Over the next couple of years, I would return with various members of my team, showing them the evidence, taking time to pray over each piece that we were able to find, and gaining more trust amongst the people there. On one of the visits, I brought Tim Moore with me. Tim has been involved in our work in Ethiopia for many years, and I wanted him to see, and feel the presence of this place. I wanted him to walk and discern what he was seeing based on the evidence in the Bible, the historical writings, and the Ethiopian manuscripts that had been given to us. After about a half an hour there, his feeling was just as overwhelming. His first reaction was, "You need to write about this. You need to tell people about this place!" Tim and I surveyed the area, trying to obtain every last item we could record, compiling the evidence, and writing each piece down in my journal of the discovery. Yes, the world needs to know. Without any doubt, that would be the next step. It was time to share this with the world, while encouraging the men who trusted me with the information to tell the truth of what they were holding.

At no point do I ever take what God has allowed me to see lightly. To me, this is an honor, a trust, and a revelation that God has led me on from the beginning, and He has laid the bread crumbs in front of me, guiding me to each and every clue, manuscript, person, and site. That first time they revealed the staff to me, was not to be forgotten. The feeling and exhilaration I felt when I first saw it, touched it, and then realized the importance of it, will never leave me. I've held the staff many times since, but to know that Matthew walked with this staff, shared the gospel of Jesus holding this staff, stood in the presence of King Bazen and Queen Candace (Mindetke) with this staff, visited with the keeper of the treasury of the Ark of the Covenant with this staff, and ultimately, turned it over to the angelic being appearing to him to perform a miracle while holding this staff, was nothing short of a humble overload and guiding light of the Lord to allow this pathway to be cleared. Without a doubt, it was time to share this discovery, the entire story, and then bring hope back to the people who have guarded this for centuries. These secrets have been passed from generation to generation and now were being revealed. These guardians have been selected, one to another, over the centuries, to stand watch over these secrets, waiting for the time when God gives them that feeling of release to the world. This wouldn't be the last time I would see or hold this staff, but it surely thrust a final piece into this magnificent puzzle.

Chapter 9

The Real Bazen, Please Stand Up

NOW IS AN EXCITING time in this discovery. This is another huge piece of the puzzle that came very unexpectedly as time progressed. Many times, as you research a discovery (or at least try to see if there is enough information about a particular piece to back it up) a curve ball is thrown into the picture that completely catches you off-guard and knocks you off the pathway you were heading down. This portion of the discovery was just that type of evidence. Before we continue forward, let's go back and review a few things. We will begin with the Bible, move onto Voragine and Foxe's manuscripts, and then place the final piece by returning to the Bible as we look further into the king and queen of Ethiopia. So the adventure begins again.

LET'S BEGIN IN THE BIBLE

First the Bible. As we see in the text, we gather that because the story of Philip in Acts chapter 8 only mentions that the Ethiopian eunuch (Bakos) served under Queen Candace (Mindetke), we can assume that her husband (the king) had passed many years earlier and she was still on the throne when this was written.

> *...and, behold, a man of Ethiopia, an eunuch of great authority under Candace queen of the Ethiopians, who had the charge of all her treasure, and had come to Jerusalem for to worship, was returning, and sitting in his chariot read Esaias the prophet*. (Acts 8:27)

The Scripture specifically points out Candace, with no mention of her husband, the king. And the mention of the Ethiopian eunuch, or as we have established as more of a general, or a likely priest, goes on to show a set-up of importance to the housing of what we now know to be the Ark of the Covenant. Again, this gives us the perception that the husband of Candace has passed and is no longer a ruling party in the kingdom.

VORAGINE AND FOXE'S WRITINGS

As we now reflect back on the text from Voragine's *The Legenda Aurea* (or *Legenda Sanctorum*, or *Golden Legend*), and Foxe's *Actes and Monuments* (or *Foxe's Book of Martyrs*), we can piece together the king and his whereabouts in this story. Looking closer, Foxe only gives a glimpse into Matthew's work in Ethiopia while Voragine has a more original in-depth look into the apostle in Ethiopia, and the figures involved. Voragine writes, *At this the king, whose name was Egippus, sent heralds throughout his realm, proclaiming: 'Come and see God hiding in the form of a man!'* (*Legenda Sanctorum*). This clearly tells us that the King, known in the translations as Egippus (or to the Ethiopians in its original text as Bazen), was most definitely alive at the time of Matthew's work in Ethiopia. It goes on to share that after Matthew raises the king and queen's son back to life, *King Egippus was baptized with his wife and all the people (ref. 15)*. Again, this gives us reference that this king was alive at the time of these events. The writings then take a turn when they reference a new king had come into power some time later, either taking the throne by force in a battle or by an internal uprising. This was a common occurrence of those ages. This happened in European countries as well as documented happenings during the Gondar reign of kings in Ethiopia and in Roha (later Lalibela), during the kingdoms there.

Since this is very vague and we see only the mention of Candace in the Bible text, one can assume pretty clearly that somewhere within the ministry of Matthew in Ethiopia, King Bazen Egippus had died and Queen Candace (Mindetke) took over solitary rule of the Axumite Kingdom. We can assume this because the biblical text of the book of Acts was written by who is assumed to be Luke around 80 A.D., during which time, Candace would have been the last known ruler of the previous royalty. She was the one who sent the Ethiopian eunuch, or priest, to Jerusalem before Hirtacus took over, presumably by force, since Anon, the king and queen's son, was not the immediate replacement.

STEPPING BACK INTO THE BIBLE

Now we turn the table and drop way back in the biblical historical text. This is once again a Scripture that is only mentioned within the book of Matthew's Gospel, and more importantly lays out an unbelievable piece to an ever-growing puzzle. Matthew writes, once again:

> *Now when Jesus was born in Bethlehem of Judaea in the days of Herod the king, behold, there came wise men from the east to Jerusalem,...* (Matthew 2:1)

In chapter two, I spent a great deal of time discussing the birth of Christ, and especially the wise men. Again, looking at the Scripture in Matthew 2:1, we see the wise men came into Jerusalem from the east. This Scripture has been the lone piece of biblical text that, to the world, tells us that these men originated from the eastern parts of the world. In reality, the text is very accurate in exactly how it reads, and that is the wise men entered into Jerusalem from the east, which was the Jordan River Valley's entry point, most likely through the trading routes already prepared or around the mountain's high levels. Again I say, Scripture was not written to be disassembled, but to be read as a whole because it all goes together. Breaking it apart can cause great confusion, as it has done in this case.

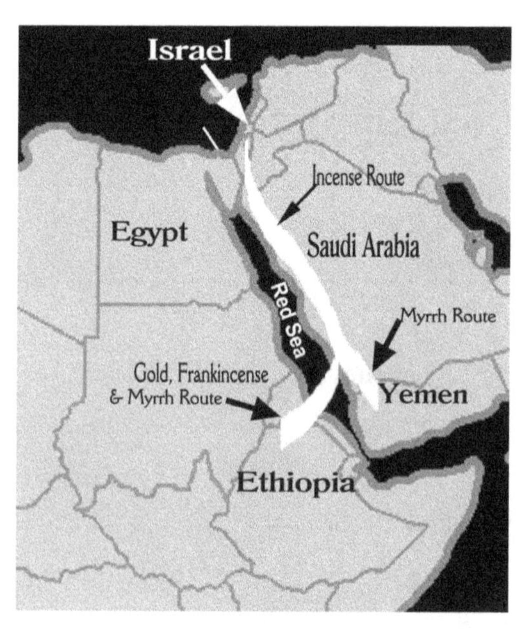

Map showing incense route from Ethiopia over the Red Sea and from Yemen

Taking a closer look at the Scriptures, they go on to give us a clear look at the direction these wise men came from.

> *Saying, Where is he that is born King of the Jews? for we have seen his star in the east, and are come to worship him.* (Matthew 2:2)

This clearly shows the wise men came from the west, making their way up the Jordan River Valley, then turning and entering Jerusalem from the eastern side, which again, was one of the traditional trading routes into the city. Although some will continue to argue this fact, I reiterate there is no way they can be from the east and see the star in the east. Many have said this is a contradicting verse from the Bible. First of all, the Bible is not contradicting, and secondly, the reader and the interpreter needs to read it as it is and not add to the text. This goes back to what I explained earlier in these writings. With the thought that this is contradicting, modern-day translations have simply removed the direction they saw the star and replaced it with wording to say they saw the star rise into the sky. This seems to be a quick fix rather than understanding the true meaning of the writings in the first place. The fact is, the wise men came in from the west, moved up toward Jerusalem, turned eastward through the mountain pass trading route, and came into the city, that is if we believe the truth of the Bible as a complete work of God by looking at the Old Testament prophecy.

Incense route of wise men coming in from the eastern side of Jerusalem.

We see even more truth to this when we refer back to the Bible. We hear well-versed people in the Word of God say there is nothing more written about the wise men, other than what we see written in Matthew's account. That is far from true. The Old Testament gives us a clear glimpse of where these men came from.

Before we look at this, let's look closer at who these men were? Over time, there have been many guesses, including the theory they were magi, or magicians, because of a mis-reading of translation that has multiple possibilities. This has led us to a wide array of thoughts on this matter, and one that is many times misused. These were men of great authority. How do we know this? First of all, King Herod would not just welcome into his realm simple magicians from a faraway land. Nor would he have *gathered all the chief priests and scribes of the people together, he demanded of them where Christ should be born*, as we read in Matthew 2:4. A king of the stature of Herod would not have put so much attention on simple magicians, or sorcerers, because they could be found all throughout the inhabited world with the great multitude of pagan beliefs at this time in history.

It's also discussed how the wise men knew to look for a star to follow. I've heard some say the book of Numbers may have been their guide, assuming these men had access to the Jewish writings.

> *...there shall come a Star out of Jacob, and a Sceptre shall rise out of Israel,* (Number 24:17)

Many others believe they may have had a dream. This is the theory I am most-likely led to believe, because these men were obviously being sent by God for a purpose. That theory is backed up by the Scripture recorded by Matthew:

> *And being warned of God in a dream that they should not return to Herod, they departed into their own country another way.* (Matthew 2:12)

This will also come into play a little further in this chapter. Until we know the full story, we can only speculate on the truth. That completeness will be coming soon as you continue to read.

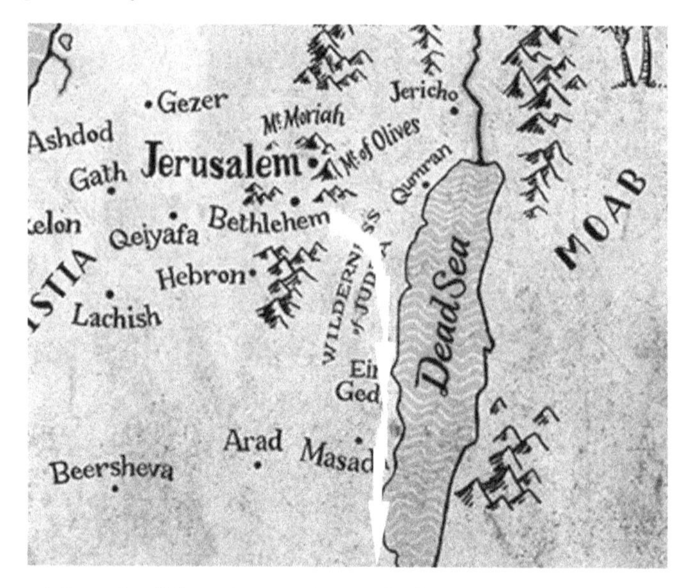

Map of the exit of the wise men's mostly likely routes along merchant trading routes out of Bethlehem

So let's take a few minutes to help you along with this journey. In most homes today, you will find a nativity scene on display during the Christmas season. My friend, and entertainer, Tony Orlando, used to perform his show in Branson, Missouri, and in his Christmas performances, Santa Claus visited him on Christmas Eve and noticed there was no nativity scene in the house. Santa then would tell him, "You cannot celebrate the true meaning of Christmas without a nativity scene to remind us of what we are truly celebrating." That, of course, was the birth of Jesus.

The nativity scene was initiated by St. Francis of Assisi in 1223, after noticing that people had lost the true meaning of Christmas and celebrations were declining. Francis asked for permission to do this from Pope Honorious III, who granted it.

St. Francis then took an ox, a donkey, and then people to pose and reenact the manger scene in a cave in the small village of Grecio, Italy. This account was recorded in the biography of a Franciscan monk named Bona-venture, who documented the event in his book, *The Life of St. Francis of Assisi*. He said that Francis would invite people to come and gaze at the manger scene while he stood and preached about "*the babe of Bethlehem*" *(ref. 36, ref. 37)*. One of the false depictions in many of the nativity scenes showing the birth of Christ is the display of the wise men gathering with the shepherds. Nowhere in Scripture does it ever put the shepherds and the wise men together. As a matter of fact, Luke's Gospel depicts the birth of Jesus, while Matthew records a time later when the wise men came into the picture. Jesus is referred to in Matthew's Gospel as now being referenced as a young child, or toddler, according to many translations of the Scripture. They are now in a home and no longer in the cave setting, or *kataluma*. And with the Bible's description of the wise men's guiding by God not to go back and tell Herod of the whereabouts of the Christ child, the king's action of having all of the children two years of age and under slaughtered gives us a clear picture that the Scripture in Luke is much earlier than the account of Matthew.

With all of this information in our minds, somewhere in time, someone added wise men to the nativity scene. Most likely it was done by the Italians as they began to offer nativities first publicly. Three wise men were added because of the thought of the three gifts (gold, frankincense, and myrrh). There was no more thought to that, other than early depictions of the wise men always appeared to have at least one darker-skinned man among them. Even most modern-day nativity scenes still depict this darker man added to the scene of the wise men. This idea was thwarted during the Renaissance period when some European artists began making all biblical figures appear to have very light-skin. This, unfortunately, has also given us an incorrect vision of what Jesus would have truly looked like. But someone knew, most likely from ancient text, that at least one if not more of the men who brought gifts to a young boy named Jesus were most definitely darker-skinned.

THE BIG CLUE FROM THE OLD TESTATMENT TELLS THE IDENTITY

It's amazing what can happen when we take the Scriptures for what they truly are. They are the guiding light to the Bible's past, present, and future, and when we view them as a whole we can see the truth appear. So let's dive into a couple of passages in the Old Testament that reveal to us directly where the wise men came from; who they were; and much more.

> *The kings of Tarshish and of the isles shall bring presents: the kings of Sheba and Seba shall offer gifts. Yea, all kings shall fall down before him: all nations shall serve him.* (Psalm 72:10-11)

First, Psalm seventy-two opens up to a piece of Scripture that will make more sense a little later, but we'll give you some geographic background so you can get a vision of the location. We see the kings of Tarshish and other locations described as isles will bring presents to the coming King of kings. Tarshish is believed to be the Phoenician port located in Spain which is mentioned throughout the Scriptures. Some claim that Carthage was once known as Tarshish, but regardless, we know this Phoenician Port of Tarshish brought Solomon great wealth during his reign as king. Again remember for later, kings from this area came to bring presents.

Now, let's take a look at the second verse (11) as it speaks of *the kings of Sheba and Seba shall offer gifts*. First, we have to look at a few facts about the area this is referencing and what they were carrying. Let's begin with the latter part of this first. It says these kings will be bringing *gifts* which can be directly paralleled with Matthew chapter 2:11 when the apostle describes the wise men *presented unto him gifts*. As for the locations this Scripture in the Psalm speaks of, we will go onto another book in the Bible to clarify this. There is so much more to be found about these men as we will see in the Prophet Isaiah's writings, which add even more excitement to the fullness of God's roadmap to reveal His mysteries when it comes to the wise men.

> *Arise, shine; for thy light is come, and the glory of the LORD is risen upon thee. For, behold, the darkness shall cover the earth, and gross darkness the people: but the LORD shall arise upon thee, and his glory shall be seen upon thee. And the Gentiles shall come to thy light, and kings to the brightness of thy rising. Lift up thine eyes round about, and see: all they gather themselves together, they come to thee: thy sons shall come from far, and thy daughters shall be nursed at thy side. Then thou shalt see, and flow together, and thine heart shall fear, and be enlarged; because the abundance of the sea shall be converted unto thee, the forces of the Gentiles shall come unto thee. The multitude of camels shall cover thee, the dromedaries of Midian and Ephah; all they from Sheba shall come: they shall bring gold and incense; and they shall shew forth the praises of the LORD.* (Isaiah 60:1-6)

This Scripture is exciting and without any doubt it gives a closer look at many of the early traditions that we still follow today, but are never talked about. Let's set this up. We must first understand these and several of the previous Scriptures in Isaiah are speaking of the coming Messiah. You'll remember the Ethiopian eunuch that Philip met was reading Isaiah fifty-three describing the Messiah and the events that were coming to identify Him. So, this continues to give us more of the coming story of the Son of God.

Here, it first speaks of the Messiah as the Light and the gentiles will gather (as well as the Jews) which could be showing us the announcement of Jesus' coming, especially to the shepherds who were called to come by the angel in the night sky, as seen in Luke chapter two. But then we see the revelation, *and kings to the brightness of thy rising*. This is a clear description of the star that rose in the sky which the wise men were following as their guide, as Matthew spoke of in his Gospel. In this incredible prophecy of the coming Jesus, and the light that guided the wise men, it directly is showing the events to come.

Now things really get exciting. We always have seen the wise men riding camels, which is most likely an assumption because this animal has always been a primary source of travel over the long hot journeys through the arid land. Here the Scripture tells us of the *dromedaries* (the one humped camel) will bring these men from the lands of Midian, Ephah, and Sheba and they will be carrying gifts of gold and incense (frankincense and myrrh). This is telling us the kings will be carrying the exact gifts presented to Jesus, as mentioned in Matthew two, and they will be on camels. It also answers the question of whether it was one, two, or three men bringing these gifts. Here, clearly, we see at least three men would travel bearing gifts from three different kingdoms.

But there is more as this unbelievable evidence describing where the wise men (or kings) came from begins to comes together to tie a secure knot of truth. This biblical text tells us much more and brings an exciting confirmation to this story. The lands of

Midian were believed to be located (in the later Old Testament) in the modern-day Southern Jordan region and into Saudi Arabia. There is now great evidence though that the Midianites migrated to this area but originated near modern-day Northern Ethiopia. We see in the Scriptures that Moses married a woman named Zipporah (Exodus 2:21), and then we learn in Numbers 12:1 that she was from Ethiopia. Within this Scripture God is angered with Aaron and Miriam, Moses' siblings, for their judgement of Moses for marrying an Ethiopian woman, most-likely because of color. As a matter of fact, the Scripture tells us she is Ethiopian twice in the same verse and we know she was from Midian. We'll leave that for another project but it's intriguing to the evidence of where the gifts Isaiah speaks of came from match the land of Ethiopia perfectly, or in today's Africa the entire region of the Great Rift Valley. Now, moving onto Ephah, which was most-likely a city and kingdom found in Saudi Arabia, and possibly even towards Yemen which would match the over-all evidence as well. And then it speaks of the lands of Sheba which some will claim were in Yemen, but there is overwhelming evidence to this particular region being located in Northern Ethiopia. There have been incredible archeological finds to the Land of Sheba being located in Ethiopia, including what is believed to be gold mines of Makeda (the Queen of Sheba), her palace remains, battle ground markings, and much more. Of course we also know much earlier in time, the Queen of Sheba delivered many of these same valuable gifts to King Solomon as referenced by Jesus in the Scriptures, and only presented by Matthew.

> *The queen of the south shall rise up in the judgment with this generation, and shall condemn it: for she came from the uttermost parts of the earth to hear the wisdom of Solomon; and, behold, a greater than Solomon is here.* (Matthew 12:42)

First, we see the reference to the Queen of Sheba as Jesus describes this woman as *the queen of the south*. We know according to the Scriptures she brought gifts of incredible value to Solomon and seduced him with her beauty and wealth. Then we find out she was from *the uttermost parts of the earth*, which is believed to be the farthest region mentioned within the Scriptures and the most undeveloped parts of the known old world at the time of the writings. To understand this better you must understand the gifts of gold, frankincense, and myrrh are found in greater quantities in this area of the Great Rift Valley, and especially in Ethiopia, than any other place on earth, which is also the land being reference when we speak of the *uttermost*. Gold and frankincense are especially found in huge amounts throughout Ethiopia (and throughout Africa's Great Rift Valley) and it's believed that's where the Queen of Sheba (known as Makeda) gained all of her wealth. With that said, this Scripture shows Jesus introducing Himself as *greater than Solomon*, which makes Him a worthy recipient to the gifts, but to this Scripture His knowledge, and the opportunities coming through Him, are far greater than Solomon could have ever dreamed of.

These Scriptures give us a direct link to understanding who the wise men (the kings) were, the connections for the gifts they presented, how many there were, where they traveled from, and how they traveled. This also gives us a confirmation of the routes these men took to arrive in Jerusalem, through incense trading routes which originate from as far away as Yemen, and across the Red Sea in Ethiopia, which geographically all would be south, and most importantly, westward from Jerusalem and Bethlehem. Again, they were known as trading, or more importantly as incense routes, specializing in the gold and valuable incense throughout the old-world merchants during the times of these biblical events.

THE REAL KING BAZEN EGIPPUS APPEARS

Large stelae marking the tomb of King Bazen in Axum, Ethiopia

In the 1960's on a hillside in the northern part of the city of Axum, Ethiopia after a day of rain in the summer, parts of a hillside gave way, exposing a rounded stone structure protruding from the ground. Not much attention was paid to the stone, because a great deal had been found and much was being worked upon at this time. Thus, the Ethiopians simply let it go for the time being. The next rainy season, the dirt from the hillside gave way more, exposing the stone of the hill, but also more of the great pillar that simply peeked from within the dirt the previous year. The history of the Ethiopians includes great obelisks that mark tombs of important individuals such as kings and queens found near the St. Mary of Zion church, where the Ark of the Covenant is said to be housed. This particular obelisk was found about a half mile westward from the church of the Ark and just off the main road and standing nearly seventeen feet tall.

Archeologists were called in, and they began to excavate the sight, uncovering a great tomb of one of Ethiopia's Axumite kings. The main tomb is surrounded by other tombs on the property.

Looking down into the tomb of King Bazen

Specifically focusing on the main tomb chamber, there are large cuts into the rock to house either bodies of servants or the treasures of the king as you make your way down the stairs. At the bottom of the main chamber were four tombs cut. Two standard size tombs and one larger double tomb to the northern side of the chamber, and another standard size tomb on the southern side of the chamber, opposite of the double tomb. When archeologists found this tomb they were quick to see that at some point in time it had been looted. All of the treasures of the king and queen had been taken - or had they? Yes, the precious items such as gold and gems were all gone, but the true treasure was in the pottery of this great king. What looters are least interested in would be the pottery housed in the chamber. Other than a few records kept by historians over the centuries, the history of the royal family had been etched into pieces of the pottery fragments in order for the world to be able to keep their story alive. This was a common practice at the time, but that's far from what the looters were interested in. But in the archeological world, this is the true treasure.

Looking into King Bazen's tomb with area for Queen Candace in raised area to the left in tomb.

This etched history shows the large tomb was the final resting place of the king, and the left raised area in his tomb was for his wife, the queen. Directly across from them was the single tomb of their trusted servant. The two tombs next to the king and queen's tomb are somewhat conflicting since the pottery fragments were not complete, and any writings never mentioned the meaning of them, thus there was no way to resolve who they were. What was found is the king that was laid in this tomb was named Bazen, and next to him was his wife, Queen Candace (Mindetke). Across from them was their trusted servant, the Ethiopian eunuch (Bakos), the high priest or general.

Within King Bazen's chamber is the tomb of the Ethiopian eunuch (Bakos).

Many believe the other two tombs within the main chamber were cut for their children, Anon and Ephigenia, while others have come to another conclusion. What would that conclusion be? To answer this, we must first finish this story.

KING BAZEN, PLEASE STAND UP

So, what was written in this pottery and in the history archives of the Ethiopian kings about this man? What could be so special, that we have held this piece of information off until the later parts of this writing? What is King Bazen hiding within the stone walls of his tomb? Well, let's share his story more in-depth to open your eyes to expose this incredible man.

What was discovered was in approximately the seventh year of King Bazen's (Egippus) reign, either he, or possibly his wife, Queen Candace (Mindetke) felt a calling from a god above (not necessarily known at the time as Jehovah God since there was confusion throughout this land with the many pagan teachings), that the king was to collect frankincense from the royal grove and deliver it to a child who would be a great King that was born in Israel. King Bazen, following the calling of a divine nature, began his travels, meeting up with two other men at the Red Sea. Eventually these three men ended up meeting nine others and then delivering his gift of frankincense to the child, thus, rejoicing for what they had seen, and traveling back to their homelands. This takes us back to the Scripture we mentioned earlier in this chapter in Psalm seventy-two that I said to remember its reference. We looked at these two verses as follows:

> *The kings of Tarshish and of the isles shall bring presents: the kings of Sheba and Seba shall offer gifts. Yea, all kings shall fall down before him: all nations shall serve him.* (Psalm 72:10-11)

It mentions *the kings of Tarshish and of the isles shall bring presents*, but then it goes on to say the important gifts worthy of the representation of the King of kings were brought by *the kings of Sheba and Seba.* The reference to Sheba is most definitely the land known as Kush (Cush), or modern-day Ethiopia, and Seba was likely the reference to the modern-day Yemen, as we saw in the Isaiah writing. Above we shared that King Bazen left on his journey to deliver the frankincense where he met with two others bearing gifts and eventually nine more men. This reference would explain this writing as recorded by the Ethiopian king and all of those he encountered along his trek. The kings from Tarshish and the other regions brought presents to the Messiah, while the kings of the west brought the gifts of the King of kings. It was also recorded that King Bazen, as he returned via the trading routes, crossed the Red Sea and entered back into Ethiopia making port somewhere in the modern-day Eritrea, which was Ethiopia at the time. He stayed around this area for a short period of time in a mountain top town, which later became known as Mount Bizen. In the 1350's there was a monastery built and called Debre Bizen, restricted to only male humans and animals. Later, the king returned to Axum, sharing what he had seen.

Mount Bizen located in the modern-day Eritrea.

So here we go. This key figure sets the foundation to the work and the mention by only Matthew in his Gospel. King Bazen (Egippus) had a sub-name, and that name was Balthazar, which he would later become better known to the world as. Thus, all nativity scenes based on the early writings depict a dark-skinned man bringing the frankincense to Christ. This is also the reason why King Herod would have allowed these men to enter into a meeting with him. They were prominent because they were kings from other nations, being led on a calling from God to deliver the royal gifts to the Christ child. Ordinary magi would not have had direct access for the most part to the king of Israel. These men did have that access, and Herod put a great effort into convincing them to return to him. This gives us a direct proof of why the Scriptures in Matthew tell us that these men came from the west, in order to present these gifts to Jesus. As a matter of fact, all of the gifts given to Jesus on that faithful day are all found in abundance within western Arabia down into Yemen, and even greater into the region of the Great Rift Valley, known as Ethiopia at the time. Finally, this is the reason we only see the Scripture of the account of the wise men to be found written by the Apostle Matthew. This is the king who delivered the frankincense to Jesus; the king who was married to Queen Candace; the king the Ethiopian eunuch (general or high priest) was guarding the Ark of the Covenant and the throne of God for; the king who would see a disciple of Jesus come to his kingdom to raise his son from the dead; the king who would receive the Salvation of the One he delivered his precious gift to; and the king who would allow this disciple of Christ named Matthew to turn to receiving and serving under the name of Jesus and the Holy Spirit.

By this time your mind is probably spinning out of control, as mine does each time I begin to go over this. I first learned of the story of Balthazar (Bazen) in 2012, but I had been putting pieces of this complex journey together since then. I never would have thought all of this would have tied together until many years later. During one visit, I was taken inside the treasury room of the Ark of the Covenant, next to the Saint Mary of Zion Church in Axum, where there are many hidden wonders. Artifacts, such as the diamond covered solid gold crowns of the kings of Ethiopia, royal robes, ancient stone inscriptions, and even the two trumpets of Moses that will be blown upon the return of the throne back to Jerusalem.

The dark wise man revealed

Etching of King Bazen (Balthazar) found in the treasury of the Ark of the Covenant in Axum, Ethiopia.

Hidden away in the back of the room amongst many old animal-skin Bibles is an etching, tucked away without much light to reveal it. I had inquired about it many times without success until one particular day. I had gained the trust of the curator of the treasury, and he pulled me to the back where the etching was located. He said to me, "This is the only image of King Bazen that exists." I quickly went to get my friend, Sisay. I grabbed him by the arm and we walked to the back of the room. He bent over to look at the face, and he relayed that this was the only face anywhere showing King Bazen's image. He said, "Mr. Jim, most don't know the story of this king; that's why they keep it located in the back. They believe the only way people will understand the king's story is if the whole story is ever told. Maybe you need to share the whole story of the black Ethiopian wise man." I replied, "Maybe I will, my friend, maybe I will someday." Well, that "someday" is now.

Chapter 10

The Apostle's Final Rest

MATTHEW HAS BEEN ONE of the most interesting figures within the Bible text, especially when it comes to his background, his lone writings (such as the description with the wise men), and his record of the Great Commission of Jesus.

> *Go ye therefore, and teach all nations, baptizing them in the name of the Father, and of the Son, and of the Holy Ghost: Teaching them to observe all things whatsoever I have commanded you: and, lo, I am with you alway, even unto the end of the world. Amen.* (Matthew 28:19-20)

When I entered Bible college, my mentor, Dr. Verlis Collins, was the man who pounded into our heads the words written in Matthew 28. He said to us, "Men, you will come to know that the Great Commission will be the foundation to everything you do in the ministry of the Lord. Whether you are preaching in a local church here in the United States, or you are the missionary serving the Lord in some foreign land, this will be your foundation because Jesus commanded it. Always know this verse, and understand it, just like Jesus sent out all of His disciples into the world so you are those same disciples. You are going into the world, preaching, teaching, and sharing the Gospel of Jesus Christ. You are baptizing them in the name of God the Father, God the Son, and of God the Holy Spirit. You are teaching them what Jesus taught the apostles, and you are to go wherever He sends you. And most of all, you are doing it in the name of Jesus. There's nothing different today than there was for these men of the Bible. You are not serving man; you are serving the Savior." That has always stuck with me. Each semester for the next three years, we would begin classes going over this Scripture and this thought. He came in to teach this, especially to embed this in our minds, and I will always be thankful for Dr. Collins and the impact he made on my life by knowing the commands of Jesus as written by His apostle, Matthew. That's why Matthew has so much impact on me when it comes to Bible teachings.

This apostle truly fulfilled that commission until his last breath. Matthew began his ministry within the local area church in Judea and then proceeded to take the Word of God unto many foreign lands. Finally, his greatest fulfillment of the Great Commission came with his decades of work in the land of Ethiopia. From this place, the apostle was challenged by many obstacles, but all resulting in confirming his faithfulness in Christ rather than hampering his efforts.

The true story of Matthew begins, as it should, in the Bible. We first learn of Matthew when Jesus confronts him and calls the future apostle to leave his life and career of sinfulness behind, to leave his world as he knows it and to follow the Master. The Bible shows no hesitation in this effort, and he simply follows. This is a true faithfulness that can't be overlooked. Jerome, one of the early Bible translators, confirms that faithfulness in his writings: "*But there is no doubt that the apostles, before they believed in Him, had seen the many signs of His power which went before Him. And of a surety, the very splendor and majesty of His hidden godhead, which shone even in His human countenance, were enough to draw them the first time they looked upon him. For if a magnet has power to attract rings and bits of iron, how much the more can the Lord of all creation draw to Himself those whom He will!*" (*ref. 38*).

The immediate attack by the sorcerers he encountered allowed the people of the kingdom to see there is One true God and no others. His faithfulness in the performing of miracles allowed the people to see God's ability, especially when he was able to raise the king's son, Anon, from the dead. This particular act brought revival to the kingdom and laid a foundation of Christ across the land.

The appearance of Jesus, or a heavenly figure, and the plunging of the rock was a two-fold effort. First, the message that all should be treated equally, no matter who you are, was clearly felt, as Jesus took us back to this Scripture, as written in Matthew's Gospel:

> *For I was an hungered, and ye gave me meat: I was thirsty, and ye gave me drink: I was a stranger, and ye took me in: Naked, and ye clothed me: I was sick, and ye visited me: I was in prison, and ye came unto me*. (Matthew 25:36)

And then, the direct power of Christ was seen when water began to flow from the rocks and life sprung forth from the desert valley. This was a direct point to the living water that Jesus was and still is, the fulfilling and refreshing way to true Salvation through only Him.

Matthew's faithfulness to the teachings of God when he stood upright against the new king and shared the Bible's stance on marriage was a strong fortress for the Word of God and the preaching that followed, and it helped seal the truth of Christ within the people. And finally, the appearance of Matthew in the fire allowed the people to fully understand they were dealing with supernatural works of the One and only God, Jehovah.

Even in Matthew's Gospel, he refers to himself as a publican, showing even more about this man. It was here we learn that the apostle was not going to drop what he was in front of the people. Voragine also writes about this fact in the *Legenda Sanctorum*. In Matthew's Gospel, chapter 10:3, he clearly refers to himself as a publican, a tax-collector, a full-fledged sinner. This sets him apart from many of the others, allowing himself to be seen as a sinner to all. In reality, a tax-collector would have been one of the most hated people; Matthew is revealing that even the worst sinner can be saved through the blood of Jesus Christ.

All of these pieces of the ministry of Matthew give us a glimpse into the real man, disciple, and evangelist. He was a man who gave up much of this world only to gain a much greater world to come.

MATTHEW'S FINAL EARTHLY JOURNEY

After Matthew's death at the hands of King Hirtacus, the people took his remains and entombed them somewhere within the boundaries of their city. At first, they wanted to hide his body from the king. Secondly, after the king's death, he would have formally had an internment into a permanent tomb, or grave. According to the Ethiopians, there is no record of the exact location of where Matthew's body had been placed, other than it was taken to a place of burial.

From this point, understandably, there is no more interaction or mention of Matthew within the history books. That is, until the Roman church came to seek the body of the evangelist in order to give him a proper and protected burial in a place of honor. This was granted by the Ethiopians, and in 954 A.D. the remains were taken into Italy. Not to Rome, but instead to a small community, at the time called Salerno. At that point this was a small church, and his remains were hidden away until many years later.

Salerno Cathedral, Italy.

According to the historians of the Duomo di Salerno, or Salerno Cathedral, the construction on the cathedral began in 1076 by builder Robert Guiscard, under the direction of Archbishop Alfano I. It wasn't completed in its first stage until 1084, when the dedication was held by Pope Gregory VII who was in Salerno in exile. But in 1081, the centerpiece of the cathedral was completed - a crypt built inside the center of the church to house the remains of Matthew and to honor him for his ministry to the world. Guiscard and the Archbishop were recorded to have been on hand to entomb the remains of the apostle in his new resting place.

Statue and display above the tomb of Matthew in Salerno, Italy.

It wasn't until 1606-1608, that an expansion of the crypt was designed and constructed inside the church by Domenico Fontana and his son, Giulio Cesare. In the 1700s, marble was added around the crypt and the magnificent vaulted ceiling of stuccoes and frescoes were painted by Belisario Corenzio with Matthew's gospel as the inspiration behind many of the scenes created. Other scenes present the history of Salerno as well as the battle the area faced at the hands of the French. Surrounding his tomb are many other servants in the church as well as martyrs of the faith, including the tombs of Pope Gregory VII and saints of the church, such as Caio, Fortunato, and Ante. Matthew's center-piece is the basis to all other areas of the church, and today it's a spectacular site to see, with two pairs of brass candelabras by Francesco Rosso placed high above the actual tomb. A double statue created by Michelangelo Naccherino also sits well above the crypt located down the stairs below. A bronze bust of the martyr sits above as well, which was sculpted in 1680 by Giovan Domenico Vinaccia.

Looking into the tomb of Matthew in Salerno Cathedral in Italy.

Descending the stairs, the elegance of the altar above turns to a circular hole with iron somewhat shaped as a mariner's compass, to guard from anyone entering. Visibly within this hole is a clear view of the tomb that was created for Matthew's remains, which are enclosed within. In all, it's a beautiful tribute (and a centerpiece of Salerno) for the man who left his world behind to follow Jesus' direction to "*go into all the world*."

Chapter 11

Matthew Comes Alive: Time-line to the Apostle

THE JOURNEY OF MATTHEW is without a doubt a story of twists and turns, miracles and deceit, faith and love. The life of the one-time tax collector turned follower of the Lord has been an incredible picture of hope, while at the same time a story of dedication until the very end. "Standing for His Faith" would be a good title to his life. With the writing of the Scriptures, God gives us that glimpse into each of those He chose to leave the world they knew behind and give their future into the hands of God. What we see in the later recorded history of Matthew is a dedicated man, willing to stand up to the wrong and spread the Word that has been written, which is always the right way.

This brings us to a time of processing, as our ministry team nurse, Kathryn Pearce, always says. It's a time to process this entire journey. To see the beginning in a new way because of what we have found in the final outcome. Matthew never backed away, never backed down, and most certainly, never preached anything other than what Jesus had taught him to share. Within this adventure into the life of Matthew, there have been many exciting events that have renewed the faith of not only those who witnessed them, but also of me – a faith to persist forward, even in the unknown struggles of life.

With all this now behind us, let's take some time to fit the puzzle together. We'll put each piece in a timeline in order for you to begin to see all of this incredible story together. The entire journey is amazing in its own right, but I will have to say, seeing this laid out in order helped show me the impact and the blessings of what has been sacrificed for the faith. Let this timeline to the apostle open your heart and eyes to what he did, sacrificed, and accomplished. Let it be a confirmation of seeing the Word come to life in the eyes of Balthazar (Bazen), Candace (Mindetke), the eunuch (Bakos), Anon, Philip, and the importance of Ethiopia.

Let's begin to piece the puzzle together for a clear timeline of history and faith.

THE PROPHECY OF THE WISE MEN

As we saw in chapter nine the wise men were prophesized well before the birth of Christ. Looking closer at this once again, we go to the Old Testament prophet Isaiah and his description of the coming Messiah and the events that would reveal Him. In the chapters prior to chapter sixty we learn what the Messiah will look like, how he will be treated, and the signs to His coming. In chapter 60 we get a great glimpse of the coming of the wise men to honor this true King of kings.

> *The multitude of camels shall cover thee, the dromedaries of Midian and Ephah; all they from Sheba shall come: they shall bring gold and incense; and they shall shew forth the praises of the LORD*. (Isaiah 60:6)

Isaiah tells us that wise men will be riding on the backs of camels. Not just any camel, but the backs of the dromedaries, or the one humped camel. Those are exactly what would have come from this part of the world at this time. Trust me, I know from personal experience as I have ridden my share of these animals over the years. The Scripture also states these dignitaries will be coming from Midian, Ephah, and Sheba. Breaking these down, we know from all archeological evidence the land of Sheba is the same place the Queen of Sheba (Makeda) came from, which is the northern portions of today's Ethiopia. Turning to Midian; we know the Midianites settled in the Saudi deserts just below Jordan, but it is also believed by many that they originally came from a mountain region in northern Ethiopia before migrating to Saudi Arabia later. Remember, Moses' wife was an Ethiopian (a Midianite) as we learned in Numbers chapter twelve. This can be traced directly to the mountains in Northern Ethiopia. Finally, Ephah is believed to be in either southern Saudi Arabia or more likely in the region of Yemen today. The gifts that are mentioned by Isaiah, gold and incense (frankincense and myrrh), are abundantly found within this part of the world, especially in the region of northern Ethiopia. We clearly see the wise men coming from the three kingdoms of Yemen, and two locations in Ethiopia riding on camels and bringing the exact gifts as we see Matthew describe in his Gospel text. This goes to show us the truth in the Scriptures along with the who, from where, on what, and the items they will be bringing? God's Word is so amazing if we'll only allow Him to pave the way and show us, and this Scripture is proof of that fact.

THE BIRTH OF JESUS

This event, recorded in Luke chapter two, gives us clear look at the events leading up to Jesus' birth and who came to visit.

> *And so it was, that, while they were there, the days were accomplished that she should be delivered. And she brought forth her firstborn son, and wrapped him in swaddling clothes, and laid him in a manger; because there was no room for them in the inn. And there were in the same country shepherds abiding in the field, keeping watch over their flock by night. And, lo, the angel of the Lord came upon them, and the glory of the Lord shone round about them: and they were sore afraid. And the angel said unto them, Fear not:*

> *for, behold, I bring you good tidings of great joy, which shall be to all people. For unto you is born this day in the city of David a Saviour, which is Christ the Lord. And this shall be a sign unto you; Ye shall find the babe wrapped in swaddling clothes, lying in a manger. And suddenly there was with the angel a multitude of the heavenly host praising God, and saying, Glory to God in the highest, and on earth peace, good will toward men. And it came to pass, as the angels were gone away from them into heaven, the shepherds said one to another, Let us now go even unto Bethlehem, and see this thing which is come to pass, which the Lord hath made known unto us. And they came with haste, and found Mary, and Joseph, and the babe lying in a manger. And when they had seen it, they made known abroad the saying which was told them concerning this child. And all they that heard it wondered at those things which were told them by the shepherds. But Mary kept all these things, and pondered them in her heart.* (Luke 2:6-19)

The Scripture not only gives us the events of the birth, but goes beyond to share even more with us. We know from Luke's singular writing of this event, the birth of Christ happened overnight, as the angel appeared to the shepherds in the fields during the night. We also notice that these men, simply tending to their flocks, were fearful of the angel and the multitudes of others that surrounded them. I would have too, wouldn't you? Think about it. A calm night as usual, then out of nowhere the night sky is filled with angelic beings, telling me not to fear. As I laugh, I can't help but see in my mind the faces of these shepherds as this event took shape.

We see the shepherds were much like the disciples; they dropped everything and then proceeded to some type of dwelling place in Bethlehem. They left immediately, as the Scripture in Luke tells us. There was no hesitation on their part, as we see in verse 15, *Let us go now*. And then Luke also shares this thought again in verse 16, *And they came with haste*. This means, by definition, that these shepherds came urgently and quickly. Wouldn't you? Think about it, the sky is full of angelic beings telling you to go see this child. I would leave right now. All joking aside, this Scripture is written in a way to show us that when God calls us to action, we are not to tarry, but to arise and go now.

BALTHAZAR APPEARS, JOINING THE OTHER WISE MEN

As we stated earlier in these writings, Matthew is the only disciple to write anything in the Bible about these men of stature and wealth. Why did he write this? It's obvious now he had a direct tie to one of these men, King Bazen Egippus Balthazar, who confirmed the story that God had inspired him to write. We must also assume that at some time during His discipleship of His followers, Jesus would have known who would be chosen to have the Word breathed into them in order to write and record the Gospels and would have prepared them for their additional tasks ahead.

The words written and discovered inside the tomb of Balthazar in Axum, Ethiopia, along with other historical recorded text about him, state either King Balthazar (Bazen), or his wife, Candace (Mindetke), felt a calling from a god above. Now we know that this calling was from the One true God, Jehovah, that was guiding the dream for him to take the frankincense from his personal grove to a child who would be King in Israel. Frankincense is found in large amounts in the region in and around Ethiopia, which confirms this even further. As a matter of fact, gold and myrrh are also found in this part of Africa in huge amounts. The Ethiopian historical text goes on to share, along this path, Balthazar would meet up with two other men bearing gifts, and eventually nine others would make the journey with them.

As we saw in chapter nine the wise men were prophesized well before the birth of Christ. Within the text of Psalm 72:10-11 we see clearly a few things opened up to us that will give a clearer picture of the Ethiopian writings sharing that Balthazar will meet with two men and eventually nine more, making twelve total kings visiting the Christ child. In the Psalm, which is describing the coming Messiah once again, it states:

> *The kings of Tarshish and of the isles shall bring presents: the kings of Sheba and Seba shall offer*

> *gifts. Yea, all kings shall fall down before him: all nations shall serve him.* (Psalm 72:10-11)

This clearly shows us the kings of Tarshish, and of some undescribed isles (which could also translate to countries) would come to visit the Messiah and bring Him presents. It then states the kings of Sheba and Seba will bring Him gifts which parallels with the gifts Matthew describes the wise men bringing to Jesus. We know that Sheba was the land in which the Queen of Sheba would come from. Although many have twisted this belief of her origination, her name was Makeda (or Saba) and according to all archeological evidence, she came from what is today northern Ethiopia. That would show the direction of the wise men matching, based on the evidence we have from the Bible Scriptures. We can confirm this also by what Jesus shared in Matthew 12:42, when He describes Sheba as the *Queen of the South*, and coming from the *uttermost*. The uttermost, or the farthest regions, has always been described as Ethiopia, which was looked upon by many in the theological world as a place of undeveloped civilizations. They were not as developed as the rest of the known world at the time, but that may have been God's plan in order to be the place He would hide many of His secrets and protect His early foundations of the Bible. Without a doubt it's also a land of great wealth in their natural resources which has sustained them for centuries and centuries.

Again, as we did earlier, we will break this down so you can directly see the long-taught misconception of where these men came from, simply by taking the Scripture out of context. Let's begin with the entire passage in Matthew 2:

> *Now when Jesus was born in Bethlehem of Judaea in the days of Herod the king, behold, there came wise men from the east to Jerusalem, Saying, Where is he that is born King of the Jews? for we have seen his star in the east, and are come to worship him. When Herod the king had heard these things, he was troubled, and all Jerusalem with him. And when he had gathered all the chief priests and scribes of the people together, he commanded of them where Christ should be born. And they said unto him, In Bethlehem of Judaea: for thus it is written by the prophet, And thou Bethlehem, in the land of Juda, art not the least among the princes of Juda: for out of thee shall come a Governor, that shall rule my people Israel. Then Herod, when he had privily called the wise men, inquired of them diligently what time the star appeared. And he sent them to Bethlehem, and said, Go and search diligently for the young child; and when ye have found him, bring me word again, that I may come and worship him also. When they had heard the king, they departed; and, lo, the star, which they saw in the east, went before them, till it came and stood over where*

> *the young child was. When they saw the star, they rejoiced with exceeding great joy. And when they were come into the house, they saw the young child with Mary his mother, and fell down, and worshipped him: and when they had opened their treasures, they presented unto him gifts; gold, and frankincense, and myrrh. And being warned of God in a dream that they should not return to Herod, they departed into their own country another way. And when they were departed, behold, the angel of the Lord appeareth to Joseph in a dream, saying, Arise, and take the young child and his mother, and flee into Egypt, and be thou there until I bring thee word: for Herod will seek the young child to destroy him.* (Matthew 2:1-13)

Immediately, Matthew chapter two tells us that Herod was the ruling king at the time and the route these men of stature took when entering the city of Jerusalem.

THE EAST INTO JERUSALEM

Because of the mountainous terrain surrounding the city of Jerusalem, there were ancient trading routes leading into the city. These were pathways for the merchants and the people to use when going from place to place, such as coming from the Mediterranean Sea from the northwest, from the rift of the Jordan River Valley from the northeastern areas, or in this case, from the southern Jordan River Valley from the southwest. These were known trading routes, and depending on where you would have been coming from, the route that would have been used to arrive in the city.

In this case, the Bible directly says the wise men entered into the city of Jerusalem from the east, which tells us they had traveled up the Jordan River Valley, mostly likely all the way to the Dead Sea, then through the trading route bypass through the mountains, and arriving into Jerusalem: *there came wise men from the east to Jerusalem* (vs.1). Nowhere does this say they came from the eastern parts of the known world at this point. This Scripture simply tells us they had arrived from the east into the city.

THEY SAW HIS STAR IN THE EAST

Immediately, in the second verse of Matthew two, we see these wise men telling us they had seen the star in the east, which directly gives us more of a location of where they had come from, *for we have seen his star in the east, and are come to worship him* (vs. 2). If they saw the star in the east, then it would not have been possible for them to have also come from the eastern known world, as most of the world will falsely argue. This is simply an assumption of Scripture by taking the first directional sign we see and applying it as fact regarding their homeland. This fact is confirmed again in verse 9 as they tell of the star they saw in the east. Again, many translations have simply changed this reading to seeing the star rise in the sky without any direction in order to side-step any thought of contradiction. It's not contradiction if you know the Scriptures for the truth of what was truly written. Many may also argue they were saying they saw the star move in the east at this point in the Scripture, which is not true, because Bethlehem is south of Jerusalem. They were simply restating the direction they had seen this star in the sky for quite some time, and it was guiding them to where Jesus would be found.

NOT SIMPLE SORCERERS OR ASTROLOGERS

Over time, many have taken the multiple possible translations of the words of the wise men and have chosen the one that matched their reference to seeing a star and put them together, coming up with the best match in a secular point of thinking as men of astrology or even sorcery. This would have been the best description by using a secular position. If we step away from the world and use a divine position, you begin to see a clearer picture. By no means were these men simply astrologers or sorcerers. How can we determine this? First, we must understand that there were sorcerers and astrologers all throughout every kingdom during this paganistic time period. Herod had his own that surrounded him, along with the chief priests and scribes as described in verse four.

These were men of wealth and power, and that well-known fact paved the way for them to have an audience with King Herod.

They were wise men, many times used in the descriptions of kings, even in the Scriptures. Their presence was so great, Herod called for the chief priests and scribes to assemble to give him direction of the prophecy of the great King that had been born - the same King the wise men had said in verse two was the reason they had made this long journey in the first place. This claim simply coming from some traveling astrologers would have had no bearing on a king of Herod's status. But coming from men of wealth and power (and even from kings) would change the opportunity to have an audience in the presence of Herod. After hearing the prophecy from the chief priests and scribes, Herod called the wise men back to him in verses seven and eight, and attempted to lie his way into having them come back to share where Jesus was located so he could worship Him as well.

THE GIFTS FROM KINGS TO THE KING OF KINGS

Once the wise men departed from Herod's palace, they began their trek to follow the star, which was now hovering over the place of where the child was located. Scriptures tell us in verse 10, *When they saw the star, they rejoiced with exceeding great joy*, that when these great men arrived at the place of the King they were searching for, they broke out in celebration and probably some tears as the journey had to have been long. By this time, Jesus was in a house, and the reference to a child in verse 11 tells us that it was some time since His birth, because the reference to a child means he was more of a toddler by this time. And the decree of Herod in verse 16 tells us that he ordered that all the children from two years of age and younger, were to be killed. He was basing this from the time he had met with the wise men, after realizing that they were not coming back.

> *Then Herod, when he saw that he was mocked of the wise men, was exceeding wroth, and sent forth, and slew all the children that were in Bethlehem, and in all the coasts thereof, from two years old and under, according to the time which he had diligently inquired of the wise men.* (Matthew 2:16)

When the wise men approached Jesus, they simply fell on the floor in front of Him and worshipped the child. It was clear these men felt the presence of the One true King, and their long journey was fulfilled by seeing and feeling the presence of God before them.

Upon their arrival, they presented three gifts to Jesus. All of the gifts represented kingship, and were gifts of wealth; again, these would not have been carried by just anyone. We are told that gold was given to Him, which represented His Kingship. Gold was known as the precious metal of kings, and this gift was to recognize Him as the King of all kings. The myrrh was used during this time period of the Bible for embalming. It's also a soothing scent, almost hypnotic when burned. Myrrh most likely was given, being impressed upon the particular wise man that brought it, because of the suffering that Jesus would have to go through in his life to come. No doubt, the myrrh was an unusual gift, but one that had been foretold of what was to come in the Messiah's life.

Finally, the frankincense was brought by Balthazar. According to what you read thus far, Ethiopia is a place where this precious incense can be found in a number of locations. This would match this king's calling, in order to bring the valuable item directly from his kingdom. The frankincense represented the High Priesthood of Jesus. Frankincense was burned when in prayer in the temple, and was often mixed with oil for the anointing of the priests of Israel. It was a gift of thanksgiving and a gift of praise to not only the King of kings and Lord of lords, but also the High Priest in all His glory.

One other key point about the wise men coming to Jesus: all three of the gifts brought to Him can be found in abundance in much of Ethiopia, Kenya, Sudan and the surrounding countries (which were all known as Ethiopia, or Cush, in the ancient times). There are also areas in western Yemen and other regions around the Red Sea that myrrh can be found in large amounts as well. As for the gold, it can be found in abundance in the area in and around the ancient land of Ethiopia. The facts of these gifts coming from this region matches the biblical text and justifies this unbelievable evidence from Ethiopia.

WISE MEN GUIDED BY GOD

We are finally shown in verse twelve the guidance of God upon these men. Whether it was a dream to someone else to move them to go on this journey, or a dream upon themselves, it was truly God that guided them.

> *And being warned of God in a dream that they should not return to Herod, they departed into their own country another way.* (Matthew 2:12)

This direct reference to the warning given them by God to divert from returning to Herod shows us the leading had come from God all along. This same reference is seen as Joseph is told in the next verse to depart and go into Egypt because of the wicked order of massacre that was coming from Herod.

JESUS CHOOSES HIS DISCIPLE MATTHEW

It's at this point that things begin to take shape and come together. Nearly thirty years later, Jesus began His trek across the countryside in order to hand-pick those men whom He would call His apostles, His followers, and His future evangelists, teachers, and preachers. This was a very unlikely group of men, but Jesus knew beforehand what – and whom, He was looking for.

A TAX COLLECTOR IS ASKED TO FOLLOW

Jesus makes His way into the city of Capernaum, as we see in Matthew 8:5a.

> *"And when Jesus was entered into Capernaum,"* (Matthew 8:5a)

From this point, we have placed where Matthew was serving as a tax-collector for the Roman government. Again, Jesus encounters the future apostle in the most unlikely of ways, collecting taxes, which was a hated position of the common people within the Roman empire.

> *"And as Jesus passed forth from thence, he saw a man, named Matthew, sitting at the receipt of custom: and he saith unto him, Follow me. And he arose, and followed him."* (Matthew 9:9)

> *"And after these things he went forth, and saw a publican, named Levi, sitting at the receipt of custom: and he said unto him, Follow me. And he left all, rose up, and followed him."* (Luke 5:27, 28a)

It was at this time we see Jesus' calling to Matthew to drop his current life and to follow Him into a life of ministry. He knew this was the man that would leave a world of sin and deceit and follow Him until the end.

HIS EXCITEMENT OF HIS CALLING IS SHOWN

There was no doubt Matthew was extremely excited about the feeling of peace he experienced when Jesus called, but also to celebrate the new life that was before him, so he threw a party for Jesus at his home.

> *"And it came to pass, as Jesus sat at meat in the house, behold, many publicans and sinners came and sat down with him and his disciples." Matthew 9:10*

> *"And it came to pass, that, as Jesus sat at meat in his house, many publicans and sinners sat also together with Jesus and his disciples: for there were many, and they followed him." Mark 2:15*

> *"And Levi made him a great feast in his own house: and there was a great company of publicans and of others that sat down with them." Luke 5:29*

One interesting point regarding Matthew's excitement was the fact he invited more of those like him to the party. It was filled with politicians, tax collectors, and other sinful associates to this gathering. That was most likely the only people that would associate with him during this time in his life, but it was exactly who Jesus was looking for. We see within these Scriptures that those who came to the party all left knowing and following Jesus, "*for there were many, and they followed him.*" (Mark 2:15).

At this point we learn how the cowardly scribes and Pharisees confronted the other disciples concerning Jesus' mingling with Matthew's sinful friend. But it was Jesus who finalizes that accusation back to them with His reply, *I came not to call the righteous, but sinners to repentance.*" (Mark 2:17). It was time. Time for Jesus to move on, and joining Him would be unlikely candidate, Matthew (Levi), the publican.

JESUS' FINAL COMMAND

We press forward and arrive at Jesus' final appearance to the group of disciples. Jesus has returned from His resurrection, prepared the disciples for their ministry, taught them well, and now makes one final command to them all.

> *Go ye therefore, and teach all nations, baptizing them in the name of the Father, and of the Son, and of the Holy Ghost: Teaching them to observe all things whatsoever I have commanded you: and, lo, I am with you alway, even unto the end of the world. Amen.* (Matthew 28:19-20)

This Great Commission gave the disciples the order and the hope to continue forward. They had received their knowledge and education from Jesus, and they are now given the command to move on into the whole world. Ironically, once again we only see this account recorded in two of the gospels - Mark and Matthew.

THE ARRIVAL OF THE FINAL PIECE: THE HOLY SPIRIT

The Scripture in Acts chapter two is the final piece of the puzzle the disciples had been waiting on before departing into the world and sharing the Gospel of Christ.

> *And when the day of Pentecost was fully come, they were all with one accord in one place. And suddenly there came a sound from heaven as of a rushing mighty wind, and it filled all the house where they were sitting. And there appeared unto them cloven tongues like as of fire, and it sat upon each of them. And they were all filled with the Holy Ghost, and began to speak with other tongues, as the Spirit gave them utterance. And there were dwelling at Jerusalem Jews, devout men, out of every nation under heaven. Now when this was noised abroad, the multitude came together, and were confounded, because that every man heard them speak in his own language. And they were all amazed and marvelled, saying one to another, Behold, are not all these which speak Galilaeans? And how hear we every man in our own tongue,*

wherein we were born? Parthians, and Medes, and Elamites, and the dwellers in Mesopotamia, and in Judaea, and Cappadocia, in Pontus, and Asia, Phrygia, and Pamphylia, in Egypt, and in the parts of Libya about Cyrene, and strangers of Rome, Jews and proselytes, Cretes and Arabians, we do hear them speak in our tongues the wonderful works of God. And they were all amazed, and were in doubt, saying one to another, What meaneth this? Others mocking said, These men are full of new wine. But Peter, standing up with the eleven, lifted up his voice, and said unto them, Ye men of Judaea, and all ye that dwell at Jerusalem, be this known unto you, and hearken to my words: For these are not drunken, as ye suppose, seeing it is but the third hour of the day. But this is that which was spoken by the prophet Joel; And it shall come to pass in the last days, saith God, I will pour out of my Spirit upon all flesh: and your sons and your daughters shall prophesy and your young men shall see visions, and your old men shall dream dreams: And on my servants and on my handmaidens I will pour out in those days of my Spirit; and they shall prophesy: And I will shew wonders in heaven above, and signs in the earth beneath; blood, and fire, and vapour of smoke: The sun shall be turned into darkness, and the

> *moon into blood, before that great and notable day of the Lord come: And it shall come to pass, that whosoever shall call on the name of the Lord shall be saved.* (Acts 2:1-21)

It was now complete, and the time had come for the disciples to no longer join together and travel, but to spread out across the known world and preach the Gospel to everyone. This would be the beginning of the ministry of each of the disciples of Christ. This would be the beginning of a courageous journey, through thick and thin, to bring the people of the world to repentance and to the saving opportunity and power that awaits by receiving Jesus Christ into their lives as Savior.

PHILIP CALLED TO THE ETHIOPIAN MAN OF AUTHORITY

Luke records in Acts chapter eight a full account of Philip's calling by God to go to a desert in Gaza, to encounter an Ethiopian man. This great amount of Scripture tells us so very much about the disciple, the man he encountered, what the man was doing, why he was doing it, and who the man worked for. Of course, ultimately, the purpose was for Philip to lead this man to Jesus, and finally he was baptized.

WHO WAS PHILIP SEEKING?

As we now know this Scripture is important because it showed this great Bible text of Philip spreading the Gospel to a faraway nation through a man of importance. It's not the simple fact that Philip introduced the Ethiopian to Jesus' salvation, but it also showed the Gospel being sent well into the uttermost. One other point we must realize is that this Scripture also gives us a glimpse to other important facts about the eunuch which ties him into other portions of the Bible. Breaking this down, let's first look at Acts eight where it describes who Philip was going to Gaza to encounter.

> *And he arose and went: and, behold, a man of Ethiopia, an eunuch of great authority...* (Act 8:27)

Philip encounters an Ethiopian man, a eunuch as described here, but someone with great authority. Again, as we researched into the ancient Ethiopian texts, we discovered that before translations into the European text the eunuch had the name of Bakos. When the Scriptures were translated, the man of great authority was simplified to a servant eunuch, but the inclusion of his importance gives us another view. This, again, would have been a general (as given in the Ethiopian writings) or a high priest, which is most likely, because of what was in the treasury of the Ethiopians.

WHAT WAS THIS ETHIOPIAN THE GUARDIAN OF?

As we saw in chapter five and as recorded in great detail in my book *Expedition Ark of the Covenant (ref. 20)*, the Ethiopians were in possession of the great Ark of the Covenant at this time in history. As a matter of fact, it had been there for quite some time. When this event happened in Acts chapter eight, the Ark would have been housed in a tabernacle on Tana Kirkos Island, at the head waters of the Nile River. There would be no other reason for the Scripture in Acts to mention the Ethiopian "*had the charge of all her treasure*" (verse 27). This man was not your everyday eunuch, but he was a man of high authority in the kingdom in Ethiopia.

WHO DID THE ETHIOPIAN SEEK?

The Ethiopian that Philip had encountered was serving under the queen of the Ethiopians. There is not much written of Queen Candace other than what is written in the history of the Ethiopians which gives her the name in the ancient texts as Queen Mindetke. We learn later of her possible significant role with the dream either she or her husband, Balthazar, had to take frankincense to the Christ child. But for now, we know she led the country of Ethiopia, at least when this biblical text was completed, but earlier she had sent her high priest, or general, who kept watch over her treasury, which contained the Ark of the Covenant out on an assignment.

> *...had come to Jerusalem for to worship, Was returning, and sitting in his chariot read Esaias the prophet.* (Acts 8:27c-28)

This man of authority from Ethiopia had been sent to Jerusalem, but why? We learn he was sitting in a chariot, handling and reading a cumbersome scroll of the book of the prophet Isaiah. Again, we see his importance because of Luke's description the man is in a chariot. The Scripture he was reading was Isaiah chapter 53. These verses were the description of the Messiah, and the Ethiopian went on to tell Philip he was confused by what he is reading, based on the disciple's question to him, asking if he understood what he was reading. The Ethiopian's response was:

> *How can I, except some man should guide me?* (Acts 8:31)

From this point, Philip joined the man in the large chariot; he talked with him, ultimately leading him to Christ, and then going into the water to be baptized. But there is more confirming information to the reasons why the eunuch had come so far. During this time, Jerusalem would have been in much despair. Jesus was gone, people were confused, the Romans were still ruling, and the Ethiopian was bewildered. Knowing what was in the treasury of Queen Candace at this time (the Ark of the Covenant) gives us more of the overall purpose of the eunuch. It's evident his journey, most likely as a high priest, was to determine if the Messiah had returned, and if it was time for the Ethiopians to begin the journey of the Ark of the Covenant, and more importantly, the Throne of God, back to the One Who will next sit upon that coveted seat.

In all of the confusion in Jerusalem and all over the land, the Ethiopian man (named Bakos) turns around and begins his journey back home, which means he was traveling along the northwest coast above Israel, along the Mediterranean Sea, and back to Ethiopia through Egypt. God was not finished with this man because of his encounters later. Thus, Philip was sent to lead him to Christ, ultimately preparing the road for a kingdom to receive that same Salvation.

APOSTLE'S MINISTRY BEGINS

As the disciples of Christ began their earthly ministry throughout the known world, there are many avenues of manuscripts that bring us their stories, whether they were first recorded by individuals traveling with the disciples, or hearing those sent to preach in person. Either way, those nearer to the times of these events taking place were able to record many of the happenings of the apostles. But there is still much, I believe, that is still undiscovered. Irenaeus records Matthew began his preaching nearer to home, as many others did, as he spent a great deal of time with the Jews in Judea.

From this point, upon the completion of his time in Judea, as confirmed by Clement, the apostle spreads his wings into the world as commanded by Jesus and travels to Parthia (modern day Iran), and then throughout that region, including Media, Syria, Persia, and mostly likely many other areas, before moving onward to Egypt. Before heading for Egypt, Matthew was recorded to have preached in the same area as Philip the deacon, and the assumption comes into play that these men discussed events and the future plans of each. That is obvious from the events still to come. After entering Egypt, preaching and sharing the gospel there, he moved southward into Ethiopia.

After studying this information, the manuscripts, and the many writings throughout the ages that have been written, there was never a clear definition of the timetable of his ministry up until this point. All we know is that Matthew was well on his journey to completing his wonderful work in serving Christ as he was commanded to do.

THE APOSTLE MEETS THE EUNUCH

After departing Egypt, Matthew made the trek southward into the uttermost land of Ethiopia. From the collection of manuscripts accumulated by men such as Clement, Papias, Eusebius, Jocobus de Voragine, John Foxe, and others, this story of the Apostle Matthew becomes one of great faith. John Foxe writes in *Actes and Monuments* the names of some of the men who recorded Matthew's history as it was happening, such as Julius Africanus (also known as Abdias), Vincentius, Perionius, and others. These names we may not know throughout biblical history, but men who witnessed or took notes from those who did, much the same as Luke did in his gospel by interviewing those who witnessed the events of Christ as they happened.

Matthew's first encounter, which we must assume was more of a search, was with the Ethiopian eunuch (high priest or general named Bakos). This, again, takes us back to Matthew's time preaching and encountering his friend, Philip, along the journey. Philip, without any doubt, had a hand in this meeting of these two men. From this point, Matthew is taken by the treasury keeper to the kingdom from which he came. That would be known as the Axumite Kingdom in modern-day northern Ethiopia, just miles from the Eritrean border. The apostle is undoubtedly introduced to King Bazen Balthazar Egippus, and his wife, Queen Candace - Mindetke. Matthew settles in the town known as Nadabah (later known as Mahibre Dego), located on the main outskirts of the city, but still in the kingdom of the king and queen.

ATTACKS OF THE SORCERERS

Immediately, resistance takes force against this area's new man of God, as Matthew is attacked by two sorcerers, who used pagan idol worship to gain the trust of the people. The apostle preached the Word of God to these men, Zaroes and Arphaxat, and the people turned against the sorcerers as Matthew forced the two from the village. The eunuch was amazed by Matthew's ability to speak to the people in their native tongue, and the apostle told of the gift given to him by the Holy Spirit at Pentecost.

The sorcerers return once again, this time bringing dragons in against the apostle. Again, a dragon could have been some type of living dinosaur of the time (remembering the physical word dinosaur was not created and used until after the 1840's), which was the way many early writings, including the Bible, depicted these creatures. Or they could have been monstrous Nile Crocodiles which inhabit the waterways of this region. Either way, the sorcerers brought them to Matthew in order to devour him. This time, Matthew calmed the dragons to sleep, and began to preach to these men and the people once more.

It was recorded at this point the eunuch had the men removed, never to return to this village or the kingdom again. It was also later recorded that the two sorcerers traveled far to Persia, where they encountered Simon and Jude preaching there. It's almost evident these two evil doers spoke out against them, but the two disciples vanquished the men, never hearing about them again.

A SERMON ON EDEN

After the encounter with the sorcerers, Matthew began to preach to the people again in their language. It was reported the sermon he preached was incredible and touched the people in a great way. Ironically, and not by coincidence, he preached to them on the former paradise of the Garden of Eden, which according to the historical, scientific, and biblical account, would have been located partially in that region.

Matthew, no doubt, had a hold of these people. Other than what they would have heard from King Balthazar and the Ethiopian eunuch, these people had never heard the Gospel of Christ preached in such a strong, powerful, and firsthand way. These had to be exciting times in this kingdom and around the known world as the men who served with Jesus were sharing God's Word to the world for the first time.

MIRACLE RAISING THE DEAD

During Matthew's preaching on the Garden of Eden, a great cry rang out amongst the kingdom as the king's son had died. The Ethiopian eunuch was sent to collect the apostle and to bring him to the palace of the king where the son laid motionless. As recorded many times, the apostles were performing miracles in the name of Christ, as we see Peter do outside the gate of the temple in Acts 3:6. Recorded here, Peter, in the name of Jesus, raises a lame man to walk. This was occurring throughout the known world, as the disciples of Christ were not only preaching and spreading the Word of God, but were also performing miracles, always in the midst of crowds so many could witness it. A miracle was performed in the presence of many, as the apostle commanded in the name of Jesus that the son of the king, a boy named Anon, would raise to life once again. When the boy stood, the kingdom of the king and queen erupted in joy. It spread throughout the kingdom, and the royal family received Jesus into their lives and were baptized in the name of the Lord.

This event led to an incredible revival amongst the people, as Matthew continued to share the Gospel of Christ, bringing people to the knowledge, repentance, and into the faith of Jesus as Savior into their lives.

PLUNGING OF THE STAFF

It has also been recorded over time that Jesus (or a heavenly being appearing as Him) would make appearances to assist in the ministry of the apostles. In this case, a beggar had appeared in the village in which Matthew was living and preaching. It was apparent, from the treatment the beggar received by one of the men at the church that had been erected after the raising of the king's son, that not all of those serving had taken on the belief, or kindness, of the apostle. One of the men kept refusing the beggar water but was ordered to do so by one of the priests in the church. He did so, but in a broken cup, somewhat degrading the beggar. This is where the Lord takes over.

As onlookers watched, the beggar restored the cup into a larger vessel, and filled it with water. The man was taken to Matthew, near a site where a massive rock was balancing on a small rock. The people before had believed the rock was able to balance in such a way because of the gods. The beggar removed His hood, and Matthew recognized Him as Jesus, and the disciple fell to his knees. The beggar took Matthew's silver staff into His hands and plunged it into the rock, making water flow from it, filling the desert-like valley below and bringing up a lush garden. This site can be seen in this place today.

A rock with a never-ending flow of water going down into a valley in which an unusually large lake can be found. And then hidden from site is a jungle-like oasis in the desert landscape, with tropical trees, pools, and amazing vegetation to supply the people.

NEW KING TAKES OVER THE THRONE

It was at this point in the writings that the king and his queen are both apparently deceased, and the new rule has not gone to the rightful king, Anon, the son of Balthazar and Candace. We know this because a man named Hirtacus, not of the godly teachings of Matthew, takes over the throne of the Ethiopians, presumably by force since Anon wasn't the king. The immediate action of King Hirtacus is his obsession with the former king's daughter, Ephigenia. He wanted to take her as his wife, but Matthew wanted to make sure the new king understood God's groundwork for marriage - one man, one woman, becoming one together. Matthew tells the king to follow in the footsteps of the former royal family and encourages him to attend the church service he will be holding. As Matthew begins his message to the people, the king is thrilled to hear these words Matthew speaks upon. It was an eye-opening experience for the king, according to the writings of Voragine. But his attitude changes when the apostle begins to speak directly on God's law for marriage. It was a common practice of that time for a king to have multiple wives, and even sharing within the kingdom, according to Romans and Greek morals. King

Hirtacus was not as welcoming of this message and decided Matthew had said enough.

MARTYR OF THE APOSTLE

King Hirtacus became so enraged with the one wife preaching, he left the church where Matthew was ministering. We learn the apostle continued to preach a spirited sermon and then turns his efforts toward the former king's daughter and her virgin court that surrounded her. After the new king left the church, he returned to his quarters and calls upon a swordsman, or as another translation calls the man a tormentor. Either way, he was an assassin with the orders to find and eliminate this man who stands between him and the young girl. It appears as though the swordsman waited until Matthew had finished his preaching. After this event, Matthew goes back into the church, kneels upon his knees, and goes into worship with his hands raised up to God in prayer. While the apostle is in prayer, the assassin fulfills the order of the king, takes his sword, and drives it into the back of Matthew, killing him as a martyr in his faith to Jesus Christ. The people, including the eunuch, take the remains of the apostle and place him into a tomb. Matthew served in this northern Ethiopian community for thirty-three years, laying the groundwork for generations to come and missionaries returning to the country today to help in many different ways.

ONE FINAL APPEARANCE

King Hirtacus feels as though he is free to take over the young princess, Ephigenia, and attempts to go to her quarters to take her as his wife. The people and her court protect her from the king, which again enrages him. As the writings state, he leaves, but then returns to set the quarters of the princess on fire in hopes of killing her and all those serving under her.

This is one of the most exciting pieces of this story, as Matthew appears in the fire, and somehow the flames leave the princess' quarters, and the king's palace becomes engulfed in a blaze instead. All of the king's servants were killed in the fire, and the king fell into deep despair. The king's son is then described as being filled with demons after going to Matthew's tomb, openly confessing the sins of his father. With the way the description is presented, it would appear as though it may have happened while others were gathered at the tomb honoring Matthew. Back to King Hirtacus: after his son's possession, the king became plagued with leprosy. At this time in history, there was no cure for leprosy, and the people with it were looked upon as cursed and the lowest of the population. The king decided at this point to take his own life, raising his sword, and committing suicide in the same way he had Matthew assassinated.

RIGHTFUL KING GAINS HIS THRONE

After the king's death, the people united and anointed Balthazar and Candace's son, Anon, back as the rightful heir to the throne. Anon had seen it all, from his death, to his raising by Matthew, losing his opportunity to rule by what appears to be an overtaking of the kingdom of Axum. But in the end, justice was established as Hirtacus took his life, and the young man was now the king he was supposed to be. This also brought faithfulness that Matthew had helped establish over time back into the kingdom.

FINAL RESTING PLACE

Matthew truly lived the life that Jesus spoke of in all ways. He took up the cross and followed Christ by fulfilling the Great Commission and preaching to all the world, or at least the world that he was led to serve during his lifetime. When Matthew arrived in Ethiopia, his ministry had a great impact on this land, uniting with the man Philip the Deacon encountered, and learning of the remaining pieces of the story about the wise men coming to Jesus when the Lord was a child, delivering the frankincense and honoring Him for His high priesthood.

The apostle battled the paganistic world he was surrounded with, was able to turn the kingdom to serve God, and was steadfast in his preaching of the Scriptures until the end of his life. After nearly nine hundred years, the Romans came to Ethiopia to obtain the remains of Matthew's body and deliver them to a place in which he would be honored by the world. The remains were taken to Salerno, Italy in 954, and placed in a small church for protection. It wasn't until 1081 A.D. that a beautiful cathedral was erected by Robert Guiscard, under the supervision of Archbishop Alfano I.

Today, this tomb is adorned with elaborate scrolling, paintings, statues, and honor. The actual final resting place is a crypt located down the stairs below the grand structure and seen through a viewing window. It is a beautifully built tribute to a man whom the Lord trusted to continue the work that Jesus had begun.

DISCOVERIES OF THE VILLAGE AND PAIRING BAZEN'S TOMB

It was 2012 when I first became fascinated with the possibility of the frankincense king coming from this area in Ethiopia, despite the traditional teachings I had learned over my years in church that he was from the eastern world. In 2013, I began to research the trek of Matthew, or at least the possibility of his entering into the land we know today as Ethiopia. This is not the accepted nature of how to proceed. The world of self-righteous theologians, who can't separate tradition from fact or use discernment to allow God to give the final direction and answers, will falsely direct you into other lands to keep you away from the lands of God's Word. We have decided to take the work, writings, and studies of those of the past, matching it with the Bible, and then if all parallels together, begin the research into the region to see if there is possibilities of the truths to the ancient writings. In this case, it worked out just as those in the days of the apostles, later bishops and historians, and finally writers who accumulated those early manuscripts together, had written in the works and history of the Apostle Matthew.

It wasn't until 2014 that I, with the help of my wonderful friend, Sisay, and a host of others who researched pieces of this puzzle, was able to confirm the village of Matthew fits perfectly into place. It also gave us the direction of research and discovery of the truth of the wise men, their origins, and why Matthew was the only disciple to record this event based on his relationship with the frankincense king. After years of work, the story was now in order to be shared to the world.

The full discovery line of Matthew and the Frankincense King

B = Holy Bible VF = Voragine/Foxe documentation AM = Ancient Historical Manuscripts

- Bible Prophets tell of coming Messiah as well as the wise men *(B - Isa. 60:6/Psalm 72:10-11)*
- Jesus is born in Bethlehem *(B - Luke 2)*
- Balthazar (Bazen) or Candace feel calling that the king is to take frankincense to Jesus *(AM)*
- Balthazar meets other kings and visit King Herod *(B - Matt. 2)*
- Wise men deliver gifts to Jesus, worship Him *(B - Matt. 2)*
- Holy family goes into Egypt *(B - Matt. 2)*
- Jesus meets with His Father in Ethiopia, fulfilling Isa. 7, Luke 2:40 *(AM)*
- Holy family returns home, Jesus astonishes men in the temple *(B - Luke 2:47)*
- At age of 30 Jesus begins His earthly ministry *(B - Gospels)*
- Jesus chooses & teaches His apostles and disciples *(B - Gospels)*
- Jesus dies on the cross, rises from tomb *(B - Gospels)*
- Jesus gives Great Commission *(B - Matt. 18-20)*
- Jesus ascends into Heaven, Holy Spirit comes at Pentecost *(B - Acts 2)*
- Disciples begin ministry to the world, Matthew begins in Jerusalem *(B - New Testament)*
- God sends Philip the Evangelist to Gaza, meeting Ethiopian eunuch *(B - Acts 8)*
- Eunuch (Bakos) led to Jesus, goes home celebrating *(B - Acts 8 / AM)*
- Matthew preaches in Persia with Philip *(VF)*
- Matthew goes onto Egypt, then meets in Ethiopia with the eunuch *(VF/AM)*
- Matthew raises King Bazen and Queen Candace's son from the dead *(VF/AM)*
- Matthew leads royal family to Jesus, baptizes family and most of the kingdom *(VF/AM)*
- Beggar reveals Himself to Matthew, plunging staff into rock, water flows *(AM)*
- Hirtacus forcefully takes over Ethiopian throne *(VF/AM)*
- Hirtacus has Matthew killed *(VF/AM)*
- Hirtacus cursed with leprosy, kills himself *(VF/AM)*
- Anon (Balthazar & Candace's son, boy raised by Matthew) takes over the throne *(VFAM)*
- 954 A.D. Matthew's remains taken to Salerno, Italy
- 2012 Dr. Rankin begins to research Balthazar's tomb
- 2014 Dr. Rankin discovers Nadaber (Mahibre Dego), Matthew's village
- 2015 Dr. Rankin is handed the staff of Matthew
- 2019 final pieces in the puzzle are put in place

Chapter 12

Expedition Onward!

MATTHEW'S LIFE WAS AN intriguing journey from the beginning. From the time that Jesus saw him to the finality of his internment into the tomb in Italy, this man has been an example to us in many ways. The details have been overwhelming, shocking, and inspiring throughout this journey, but with Jesus and His followers, shouldn't this be expected? The life of Christ and the entire Bible should be overwhelming to us all. It should amaze you to see God's power, miracles, and what He can do if we turn and follow Him. And then we should most definitely be inspired when we give our lives to Christ and follow the teachings that God laid before us!

Jesus many times told His disciples they will suffer for following Him. When the Messiah called upon them to drop everything, they were well aware of the risks they would endure. Thankfully, they took up the cross, as Jesus asked them to do, and followed Him. Within the covers of the Bible and the historical text that would follow, we can see the faithfulness of this special group of people called the disciples of Christ - a faithfulness that many will never understand when God calls someone still today. We see it often when a missionary family sells their home, may live on the road raising funds for a time period, and then leaves this world behind and go to an unknown land to spread the Gospel. It's the same concept, and it is truly misunderstood, because to the world it's crazy, but to God's work it's a calling – a faithful calling, the same calling as those who left everything to follow Jesus in the New Testament.

GOD OPENS DOORS

Speaking around the globe has allowed me to see the reaction and belief of people when it comes to the Scriptures and the historical writings. When I speak of other books mentioned in the Bible, some people are appalled that I would say there are other writings, while others are in awe to know there is more to the story. This goes to show that we don't truly read the detail in the Scriptures. One important but fun part to sharing these incredible discoveries to groups of people is the occasional opportunity to answer questions. I love to hear from all ages when I'm able to answer questions, and some of the most intriguing and thought-provoking questions will come from children.

Questions give us opportunities to know more, and those who don't ask already have a closed mind. They believe they know it all, which is a dangerous attitude to have. One question that I get on occasion is "Why were you chosen to find these things you speak of?" My answer is always, "Why not?" The Bible is full of individuals who were the most unlikely people to complete some of the greatest tasks asked of them, but they did it because they were willing to follow the Lord. We all are chosen to do God's work, but we all won't give up our lives to go and do that work. One of my favorite phrases in the Bible is "*Arise and go*." So many times, God calls on men and women to get up and go serve Him. Most choose to go their own way, but some will choose to obey. So the phrase "*Arise and go*" lays deep in my heart. I truly believe that if you will go and do what God is leading you to do, then you will finally be fulfilled with the life He intended you to live. He did not intend on you being miserable day in and day out. He intended our life to be an adventure; it's up to you whether you step into the adventure He has created for you, or not.

Once you're serving God in the way He meant for you, He will give you the gift of seeing Him involved in everything in your life, past, present, and future. After my wife and

I surrendered our lives to serve Him, we were able to see the events, surroundings, and happenings in our daily lives as a groundwork for something to come. It's truly incredible, but it takes devotion, discernment, dedication, and a servant's heart to make this happen. That's how these journeys became realities in finding some of the most sought-after mysteries concerning the Bible. Allowing God to have control makes life more exciting and fulfilling rather than allowing the world to dictate what we do.

ADDING TO MATTHEW'S STORY

In August of 2018 I had put most of the pieces to this puzzle together, and had been somewhat preparing my manuscript. My wife Sherri, our wonderful friend Bev Regehr, and I had journeyed back into Ethiopia for a series of meetings in various locations in the country. One of those was a visit to Lake Ziway in the Great Rift Valley of Africa. We traveled across the lake to an island that housed the Ark of the Covenant during a time of uprising in 1531. The Ark actually was hidden away on this island for some seventy years until a new church, the Saint Mary of Zion Church, was constructed and still stands today. The previous church was destroyed during an uprising, and they now felt it safe for the return.

We then made our way north to Ethiopia's capital city of Addis Ababa where we made our way into the airport for our next destination of Axum. As we boarded our flight, Asgedom joined us on his way back home to Axum after business in

Addis, while Sisay was waiting for our arrival. Our flight was delayed from Addis Ababa twice before we finally boarded and headed for Axum. As the flight progressed; shortly before arriving into Axum the pilot entered into a dark-as-night storm cloud in the small turbo-jet Bombardier Q400 plane. Lightning struck from all sides as the pilot announced that conditions in Axum were too dangerous for an immediate landing, so we would circle above or we would have to return to Addis Ababa. It was tense as the lightning flashed outside the windows, but suddenly, after thirty-four minutes of circling the pilot took a quick turn to the left and then into a sharp dive through the cloud. Within minutes we came out of the cloud to a dark runway where the pilot pulled onto the tarmac swishing from side-to-side until finally coming to a stop at the end of the asphalt. We all looked at each other shaking our heads and thankful for the Lord's safety in questionable conditions. We were so late in our arrival that Sisay had gone home but our driver was waiting for us outside and we were off to the hotel for a much-needed night of recovery. The next morning we were off to Matthew's village so I could present the opportunity for Sherri and Bev to see this incredible place for the first time.

As we entered the village of Matthew, the priest stared at us because of the women coming with us. We had never come there before with women, but it really didn't seem to bother them much. Since they follow many of the old traditions concerning women near the treasury, I couldn't help but wonder what their reaction would be but I had been there so many times over the years, the respect was apparent, and we went on in our meetings with them again. As we stood in the compound around the treasury building the priest exited with the staff of Matthew. The priest began to speak and explained that no one had been allowed to handle the staff other than the high priests in the past until they handed it to me a few years earlier. Both Sherri and Bev simply stared at the staff, amazed with what they were looking at.

We then headed outside the compound walls where I showed Sherri and Bev the foundations of the ancient dwelling places, and even some more pottery that had surfaced since the last time I was there. I pointed over to the lake in the valley, showed them the bleeding rocks of iron, and then over to the cliff to look down in the jungle. We started making our way back up to the compound of the church, but first we stopped at the large tin building that housed the great balancing stone. Both Sherri and Bev were awed by its size and how it balanced on the small stone. But then Bev reached out to touch the rock, and her eyes widened as she said, "It's like I can't take my hand off this rock." By this time Sherri had reached up. Bev went on to say, "It feels like an electrical charge going through my body, and I don't want to let go." Sherri nodded in agreement, and said, "Yes, it's just like a charge moving through it, but it's lifting, in a way." I was thankful that others felt this same feeling that I had many years earlier. It's truly amazing to see what had happened here, but to see others experience this same feeling gave me a wonderful blessing to witness the thrill in the eyes of others.

GUARDIAN FINAL REQUESTS

2013 was a pivotal year in the lives of my wife and me, and our work and study in Ethiopia. God had opened opportunities in all areas, including discernment to some of the discoveries I have written about in my previous books as well as this writing. As mentioned before in this writing, the Ark of the Covenant is believed to be in or around Axum today. The place where the most focus is put is on the Saint Mary of Zion Church, which is surrounded by a high iron fence, armed guards, and watched over by a man called the Guardian. A young boy is chosen early in life by the current Guardian and will be raised by that man to watch over, guard, pray, and wait on God's guidance in every way. When the Guardian passes away, the younger trainee steps into the role as Guardian. Once that boy crosses the fence of the church of the Ark, he loses all access to everything beyond those gates ever again. He never goes out to his family or friends, and he only has a couple of men to care and watch over him as he teaches them in the process. His life of solitude is given as service to God, to watch over the throne of God, and to prepare for the day of the return of the Ark of the Covenant and the Throne (the Mercy Seat) to Jerusalem.

It was a visit in 2013 when I arrived in Axum, Ethiopia, for a multi-purpose expedition, including some research involving the village of Matthew. Shortly after arriving at my motel room, there was a quiet knock on the door and it was my friend Sisay. He was somewhat winded when he began to tell me, "Mr. Jim, the Guardian at the church of the Ark has asked for you to come to the church to meet with him." I chuckled at first, but then realized Sisay wasn't kidding with me and was very serious in what he said. We jumped into a tuk-tuk (a small motorized taxi vehicle just large enough for a driver to steer with handle bars in the front and a tight fit for two people in the back cabin) and speedily headed down the hill to the courtyard surrounding the church of the Ark. Whenever I'm in a tuk-tuk I feel like I have to lean the opposite direction to keep these things from flipping over. We arrived as the sun was beginning to drop below the horizon but there was enough light for us to get to the church and fulfill the request of the Guardian. Rather than approaching the front gate and drawing attention, I was led by Sisay down the steps past the treasury room and to a gate below. The caretaker of the Guardian met us at the gate, a wonderful man named Zemichael Berhane. We exchanged greetings and made our way to the side door of the Saint Mary of Zion Church where there were a couple of small wooden stools sitting. We stood and talked for a few minutes and then the door leading into the church slightly creaked open. Zemichael leaned in speaking to the Guardian, and I was also able to see a young boy, the future Guardian, standing back behind him. The Guardian at first stepped back away from the door, and the three of us began to talk once again. Suddenly, the door opened slightly again and the Guardian poked his head out, looked around the perimeter, and then stepped out.

At this point I was thinking, "What am I doing here?" I really didn't know why we were meeting, but I had a respect for this man for his sacrifice. Zemichael and Sisay began to speak with the Guardian; both of them have expressed to me over the years that this man was more than what his position means to them, but he was also a spiritual father and teacher to them. They have both sat with him over the years, and he has taught them both much from the Bible, the history, and about the Ark. The Guardian finally stepped toward me, as he reached for my hand in greeting. He began to speak to me through Zemichael's translation.

Dr. Jim Rankin (left) with the Guardian of the Ark of the Covenant (in hat), along with Sisay (back to photo), and Zemichael (right) at the St. Mary of Zion Church of the Ark of the Covenant.

He said, "Mr. Jim, I am thankful you came. God spoke you were here. I want you to know I am preparing myself for to die soon, as we are preparing for the return of the throne to Jerusalem in the time to come. God says you are to begin telling the world of this, as we will. The time of Jesus is coming soon." Sisay says, "Do you understand, Mr. Jim?" I explained that I did. I then went on to ask, "When will this be?" The Guardian was quick to respond, "When God says it is time. We are preparing the objects for the return. God will tell us, and we will be ready for Him." I asked about the Ark itself, as to what it looks like, and was met with a quick response once again by the keeper of the throne of God, "Yes, yes, your Bible is accurate, look into your Bible for the answers. Tell the world we will return to Jerusalem soon, this is very important."

We continued on for a few minutes, and then he nodded to me in respect as if to say, "We were now finished", as I nodded back to him to show my appreciation. He turned, entered the doorway, closed it, and was gone. Sisay said, "Wow, that was amazing. Did you understand all of this?" My exact thought was what I spoke, "I think so." After that day, I have had regular meetings with this wonderful man many times over the years. Sometimes a simple greeting, other times some eye-opening thoughts and explanations of things to come. At least once a year it seemed he would make reference to preparing himself to pass away soon. In 2018, during our last meeting together, the Guardian said to me, "I prepare for to die soon; a younger man will have to make the journey for the throne." I was puzzled, but then started to think about that. As an older man in his physical condition, he would not to be able make such a journey if God called them today to return the throne to Jerusalem. So once again we said our goodbyes, and I left the gate and headed back out with Sisay. Little did I know this would be the last time I would meet with him.

It was in July of 2018 when I received a call from Asgedom, and then a message came in from Sisay, who I could tell was truly upset. The message was simply that the Guardian had died. I knew the relationship Sisay and Zemichael had with this man, and I was deeply saddened, even to tears, for the loss of my friend, but also for the sadness both of these men were experiencing. I first thought the news of the Guardian's passing would spread quickly in Ethiopia but I found quickly that I was wrong. After hearing the news, the next day I asked the person who takes care of our social media pages to post our condolences on the page. Within hours that post had exploded with the thousands who view our page sharing the sadness as it had not become a story yet throughout the country. We were receiving comments throughout Ethiopia, and the world, to the announcement on our social media page, which we found was the first official release of the story. Although I returned to Ethiopia the next month, it was a very quick visit for another need, so we didn't talk much about the passing of the Guardian at that time, and seemed to let it go for another visit.

It wasn't until our return in February of 2019 that I was surprised with the impact of our meetings. We were in the treasury with a team I had taken into the country to see many of these incredible finds when Zemichael came rushing into the door. As the caretaker for the Guardian, his role stayed the same, and it was so good to see him again. He was wrapped in a beautiful prayer shawl with the key points of Axum sewn into it. We embraced, both excited to see each other, but I could tell he had something to tell me. One of the men with us, Brian Palmer, was standing off to the side watching us as we talked. Zemichael went on to say, "Dr. Jim, it is so good to see you. I must tell you something. The Guard(ian) the week he died asked for us to get his royal robe and hat. He put them on, walked around the church, and then asked to have picture of him taken in it. He told me to bring the picture back to him. A few days later I brought the pictures back to him, and he said to me to make sure Mr. Jim has the picture. He said it was now time for him to die. He sat down with us and taught us some things, like he did many times before. He walked around the church and looked up in the sky. Then he looked over at us, looked back up, walked over to his chair, put his head down and was dead."

Final photograph taken of the Guardian of the Ark of the Covenant before his passing.

I stood there in shock, and then tears began to flow, as they did in the eyes of Zemichael as well. He went on to tell me he will give me the picture, and arm-in-arm we walked to see the bright red robe and hat with gold accents that the Guardian was wearing which was now placed in one of the enclosed cases in the treasury. Brian Palmer, a pastor from the U.S. who was with me, knew something important was happening as he was standing nearby and saw my tearful reaction. This request from the Guardian showed me that the meeting, the calling, and the relationship was much more than the Ark of the Covenant, but it was a friendship that has been built on trust and respect and I will never forget it. It was a confirmation for me to continue forward in the work of sharing the story of Matthew, and it made me realize once again that everyone put in your path is important, it just may be later to see how important they truly are in both lives. It somehow made me realize that it was not my interest in the Ark of the Covenant or Matthew and King Balthazar that had brought us together, but a friendship of respect that was created by God along this incredible adventure.

FINALIZING THE WISE MAN

My first visit in 2019 was a full schedule of fulfilling many pre-set appointments that needed to be accomplished. During a visit to Axum, Sisay took us to the ancient part of the city where we visited the treasury of the church of the Ark. As we sent the women into the church next door, I took the men into the front courtyard of the Saint Mary of Zion Church (an area that only men were allowed) past the front gates, and up the steps to the ancient temple. Zemichael had followed us there and opened the doors to the ancient temple to show the men the contents inside. It houses a beautiful set of paintings on the walls, the musical instruments

(left) Zemichael at the St. Mary of Zion Church of the Ark of the Covenant (behind), and (right) Sisay in traditional clothing.

for worship, an ancient Bible, and many other hidden treasures.

After our time in the temple, I had the men on our team follow me outside where the coronation throne of the kings sat, overlooking the ancient entry gate into the complex. I pointed out the lion-carved drainage spouts that have been in place for centuries and the many other engravings in the walls. It was then that Zemichael came to me and grabbed my hand. In Ethiopian culture, a man taking the hand of another man is a way to show friendship or great respect. He took my hand and pulled me over to the wall near one of the lion-down-spouts, before heading down the stone stairs. There was an ancient carving of lettering on the wall, measuring approximately four feet long. He said this was an engraving of King

Bazen (Balthazar) when he ruled and was crowned King of the Ethiopians. The many times I had stood at that same spot, but had never known what that said. I told him, "Of all the times I had been right here, I never knew this." Zemichael was quick to respond, "Could be you needed to find other things before it was time for you to know. Here it is, markings of the king you will one day write about." Little did he know I had already been writing this manuscript and was nearly finished.

Inscription of King Bazen (Balthazar) near the coronation throne and the church of the Ark of the Covenant.

It was like the last piece of the puzzle before the final "i" was dotted. As my friend the high priest in Gondar said, "If you follow God, you will find what you are looking for." Just when you believe you have all the information possible, another piece of the puzzle is put in place.

PREPARING FOR THE NEXT ADVENTURE

This amazing discovery and the story behind it has brought me even closer to the Bible than ever before. All that have heard this have wanted more, and pastors that have looked into this with me are overwhelmed. One long time pastor said, "The Bible just came more alive than ever before." I believe that's what each of our discoveries has done. Another pastor from near Chicago that has spent over sixty years in ministry said, "This is so aligned with where we are today, I've never heard anything so attractive to that in my life. This information is special." These discoveries bring you closer to the Bible and help you understand there was more that happened, and the people and places within its pages are real. They weren't fantasized figures that a story was written about, but they were alive. In the confused world we live in, this helps to bring life to those that lived, physical places to those things we read about, and a glue to bring all of this together.

One older church deacon said to me when I was in a church sharing one of my presentations, "There ain't no more written other than what's in the Bible that we need to worry about. People need faith, that's all there is to it." In a way that is true, but in the confused reality that is being pumped into the minds of people, and especially our youth today, God knew in time He would need to reveal more to bring a bonding to His Word giving a glimmer of hope for people to see it is all real. He talks to us many times about His mysteries He would one day reveal. Unfortunately, some people need to get out and realize it's not the world we grew up in, but it's a world that needs assurance and hope. I say this knowing of the many biblical discoveries that have been found over the last several years, and I know God will continue to reveal more and more in the years to come.

Within all of this you are able to see the importance of the Scriptures in a whole and not taken out of context. You see the impact of the disciples, and you can even feel their hurts, but also see their faith to continue forward. I am thankful that I was allowed to be a part of this, knowing God entrusted me enough to open the right doors to walk through. Those doors are open to us all, but we must be willing to step through them and trust in the pathways He will lead us on. Are you ready for that journey?

Just as I had finished up my research on all of this, I was traveling through the mountains in northern Ethiopia, and as we began to stop, a feeling that we needed to go on came to me. My wonderful friend Asnake said that there wasn't much ahead except for a cave church far into the mountains. I knew right then and there more was to come soon. So, when you're feeling tired, and feel like it's time to stop, press a little further and see where God takes you.

As Sherri says to me often, "*It's just the beginning of the beginning, Jim,*" I know that statement is so true in all of our lives. If we focus on the Lord and what He has intended our lives to become, then He will also reinvent each day to be exciting and new for a great adventure ahead.

Matthew has truly become more to me now than just a tax-collector turned disciple of Jesus. Matthew has become real, full of life, and a man who fought until the end to share God's Word and teachings to a hurting world. What has Matthew done for you? As I saw this man come to life, I also gained hope in seeing King Balthazar trust in a calling to go where he was needed, in order to worship the Lord in greater ways than ever. We saw him follow God's whispering, discern from returning to King Herod's lies, and then we witnessed him falling at the feet of the King of kings. Through it all I saw the Bible come to life in a greater way than ever before. I saw it merge together as one whole Word of God rather than a bunch of stories collected into a book. Within this discovery we witnessed five important figures of the Bible coming to life and more importantly, coming together for one incredible journey.

This should be a road map for each of our journeys we take. We should listen to the Lord's call, then go, discern, and worship Him along the path. Our testimony is to share what God has done for each of us to give hope to the world. The twist of the frankincense king added greatly to the story of Matthew as we saw the correction of the eastern tradition set before us while understanding how the apostle's inclusion of these wise men in his Gospel was the result of a direct relationship he had with one of those kings. This again brings the Bible to life, away from man's misunderstandings, but turning to a reality we can see truth in.

In the end, where does this bring you? Where does it bring each of us? All I can do is speak for myself, and this has brought me closer to the Jesus I serve, knowing He was the guiding light along the pathway. It also brings a little more to me. It brings me to understand there is more to life than what the world wants us to know, and when we open the small box we keep our god in, and finally allow the One true God to converse in our lives, the adventures are endless. People ask me all the time, *why do I believe there is so much being revealed about the Word than ever before*? I'm quick to respond, *because we're allowing God out of the box*. I truly believe the mysteries God speaks of are really additional physical proof that His Word is true, but to be revealed in the time it's needed to an unbelieving world.

The journey in my life is far from over; I hope you see that your greatest adventures are yet to come as well. Sherri and I always look forward to each and every day, knowing God is already there, and knowing our greatest adventures are still to be discovered. Sherri stated to me years ago as all of this began in our lives, "*God is with us, if we will allow Him to guide each of our steps*." The bags are packed, the journey is prepared, and now it's time...so let the adventures begin - again!

"God intended your life to be an adventure…it's up to you whether you step into the adventure He has prepared for you. As for me and my wife, life is a daily adventure, so let it begin!"
Dr. Jim Rankin

Reference Notes:

1. The Holy Bible, King James Version (Referenced Throughout)
2. John Foxe, *Foxe's Book of Martyrs - Actes and Monuments of these Latter and Perillous Days, touching Matters of the Church*, John Day, 1563 - later published by Wilder Publications, 2009
3. Bennedict Robinson Collected *Letters of the Reformation' and [John Strype],* Ecclesiastical History *vol I-III.*
4. John Milner, *History Civil and Ecclesiastical and Survey of Antiquities of Winchester* (1795), cited in Warren Wooden, *John Foxe* (1983), Twayne, p. 106.
5. F. L. Cross: *The Oxford Dictionary of the Christian Church (Rome – Early Christian)*, Oxford University Press, New York, 2005
6. Christian Frederic Crusé, and Henry de Valois, *The Ecclesiastical History of Eusebius Pamphilus*. London: G. Bell and Sons, 1897.

7. Andrew Kippis, Nathaniel Lardner, *Eusebius, Church History 3.24.6, The Works of Nathaniel Lardner Vol. 5*, (1838), Ball, p. 299
8. Darrell L. Bock, *Studying the Historical Jesus: A Guide to Sources and Methods (Irenaeus, Against Heresies 3.1.1)*, (2002), Baker Academic Publishing, p. 164
9. Daniel J. Harrington, *The Gospel of Matthew*, (1991), Liturgical PressX
10. Eusebius, *Ecclesiastical History*, 2.1.13
11. Boris Repschinski, *The Controversy Stories in the Gospel of Matthew* (1998), Vandenhoeck & Ruprecht, p. 14
12. James R. Edwards, *The Hebrew Gospel and the Development of the Synoptic Tradition,* (2009), Wm. B. Eerdmans Publishing, p. 18
13. Glenn F. Chesnut, (1986), The First Christian Histories: Eusebius, Socrates, Sozomen, Theodoret, and Evagrius
14. Robert Eisenman, (1997), *James the Brother of Jesus: The Key to Unlocking the Secrets of Early Christianity and the Dead Sea Scrolls*, Viking Press
15. Jacobus de Voragine, (1265-75), *Legenda Sanctorum*, (additions from *Historia tripartita et scholastica* and many other historian writers) (First edition published in 1470)
16. Ryan Granger and Helmut Ripperger, (1941), adapted translation of *The Golden Legend of Jacobus de Voragine*, Longmans, Green & Company, pp. 561-566
17. William Caxton, (1483), *Lives of the Saints compiled by Jacobus de Voragine, Archbishop of Genoa Volume Five*

18. Edward Williams B. Nicholson, *The Gospel According to the Hebrews*, (1979), p. 26
19. Answers In Genesis, (2011), *Uncovering the Real Nativity*
20. Dr. Jim Rankin, (2017), *Expedition Ark of the Covenant* (*Jesus In Ethiopia*, 2011)
21. Harriet Beecher Stowe, (1852), *Uncle Tom's Cabin*, Boston: John P. Jewitt
22. Reverend John Rankin: (1833), *Letters On American Slavery*, Original Documents – later published by Garrison & Knapp
23. David Fountain, *John Wycliffe*, Bible Translation Timeline, p. 45-47
24. A.S. Herbert, (1525-1961, 1968), *Historical Cataloque of Printed Editions of the English Bible*, British and Foreign Bible Society
25. Frederick Fyvie Bruce, (2002), *History of the Bible in English*, Cambridge: Lutterworth Press
26. KJB: The Book That Changed the World, Lion's Gate, B004K6FS5W, (2011)
27. J. Hampton Keathley, III: (2004), *The Bible: Holy Canon of Scripture*, Biblical Studies Foundation
28. Flavius Josephus: *Flavius Josephus: Translation And Commentary, Vol. 1b: Judean War* (Leiden: Brill), 2008
29. Geza Vermes: (1962, 1965, 1968, 1975, 1987, 1995, 1997, 2004), *The Complete Dead Sea Scrolls In English, Revised Edition*, Penguin Books,

30. Dr. S. C. Malan: *Vicar of Broadwindsor*, (Ethiopic Text), edited by Dr. E. Trumpp, Professor at the University of Munich
31. Sacred Texts Archives, (2010), *Book of Enoch*
32. Jerome, 492, *Lives of Illustrious Men – Papias 18*
33. John Foxe, (1563), *Actes and Monuments* (original title: *Actes and Monuments of these Latter and Perillous Days, Touching Matters of the Church*), (later titled: *Foxe's Book of Martyrs*), John Daly, publisher
34. Dr. Jim Rankin, (2016), *Guardians of the Secrets Book I*, Xulon Publishing
35. Robert Cornuke, (2005), *Relic Quest: Legend Chaser*, Tyndale House Publishers
36. Rachel Nuwer, (2012) *Smithsonian Magazine, The First Nativity Scene Was Created in 1223*, Smithsonian
37. E. Gurney Salter (translator), (1904), Saint Bonaventure's, The Life of Saint Francis Assisi, E.P. Dutton
38. R. Joseph Hoffmann, translation (1994), *Porphyry's Against the Christians: The Literary Remains*, Amherst: Prometheus Books

PHOTO CREDITS:

Chapter 1: Image 1 – *St. Matthew*, Egbert van Panderen, c1605 after Pieter de Jobe 1: Chapter 2: Image 2 – *Adoration of the Magi*, Edward Burne-Jones, tapestry; Image 3 – Ancient Israel Trading Routes (public domain); Image 4 – Wise Men Entry Into Jerusalem (public domain): Chapter 3: Image 5 -- Clement of

Rome (public domain), Papias (public domain), Polycarp (public domain), Jerome (St. Jerome by Marinus van Reymerswaele, 1490-1596), Eusebius (Engraving by Frere Andre Thevet), John Foxe (Foxe's Book of Martyrs, Actes and Monuments), Jacobus de Voragine (public domain), William Tyndale (Benson John Lossing, 1891), John Wycliffe (public domain), Martin Luther (public domain): Image 6 – Duart Castle (public domain); Image 7 – Reverend John Rankin (public domain); Image 8 – *Letters On American Slavery* (Dr. J. Rankin): Chapter 4: Image 9 – Matthew Ethiopian Monastery (AIT Stock Photo, copyrighted): Chapter 5: Image 10 – The Calling of St. Matthew, Juan de Pareja, 1606-1670, Prado Museum; Image 11 – Gondar Priest (AIT Stock Photo, copyrighted); Image 12 – Tankwa Papyrus Boat (T. Moore); Image 13 -- Tana Kirkos Pole Hole (AIT Stock Photo, copyrighted); Image 14 -- Ark Footholds (Wosson); Image 15 – Measuring Footholds (AIT Stock Photo, copyrighted); Image 16 – Bowl, Harness, Hooks, Forks) (AIT Stock Photo, copyrighted): Chapter 6: Image 17 – Philip and Eunuch (artist unknown); Chapter 7: Image 18 – St. Matthew Archbasilica of St. John Lateran, Vatican (unknown); Image 19 – Martyrdom of St. Matthew at the Instigation of King Hirtacus (Anonymous, 15th century); Image 19b – Martyrdom of St. Matthew with St. Ifigenia (altar piece of St. Matthew, 1367-70, Galleria degli Uffizi, Florence, Italy): Chapter 8: Image 20 -- Dr. Jim and Sisay (K. Pearce); Image 21 – Village Door(AIT Stock Photo, copyrighted) ; Image 22 Foundations (AIT Stock Photo, copyrighted); Image 23 – Rock field (AIT Stock Photo, copyrighted); Image 24 – Pottery Remains (A. Tazeze); Image

25 – Balancing Rock (AIT Stock Photo, copyrighted); Image 26 – Water flowing (AIT Stock Photo, copyrighted); Image 27 – Lake in Valley (A. Tazeze); Image 28 – Lush Jungle (AIT Stock Photo, copyrighted); Image 29 – Bathing pool (AIT Stock Photo, copyrighted); Image 30 – Jungle Boy (A. Tazeze); Image 31 – Crops (AIT Stock Photo, copyrighted); Image 32 – (A. Tazeze); Image 33 – Old Priest with Staff (AIT Stock Photo, copyrighted); Image 34 – Staff connection (AIT Stock Photo, copyrighted); Image 35 – Jim with Staff of Matthew (S. Tsegay); Image 36 – The Staff of Matthew (AIT Stock Photo, copyrighted): Chapter 9: Image 37 – *Adoration of the Three Kings* (by Girolamo da Santacroce, Walter's Art Museum); Image 38 – Map Incense Trading (public domain); Image 39 – Map Wise Men through Valley (public domain); Image 40 – Map Wise Men Route out of Bethlehem (public domain); Image 41 – Nativity Scene (St. Peter's Basilica 2009); Image 42 – Bazen Stelae (T. Moore); Image 43 – Into Bazen's Tomb (S. Rankin); Image 44 – Bazen's Chamber (S. Rankin); Image 45 – Eunuch's Chamber (S. Rankin); Image 46 – Mount Bizen (unknown); Image 47 – *Adoration of the Three Kings* (by Girolamo da Santacroce, Walter's Art Museum); Image 48 – Bazen's Image (unknown): Chapter 10: Image 49 – Salerno Cathedral (public domain); Image 50 – Tomb Display (public domain); Image 51 -- Matthew crypt (Duomo di Solerno, Parrocchia SS. Matteo e Gregoria Magno): Chapter 12: Image 52 – Expedition photo (various, Rankin); Chapter 53 – Jim with Guardian (K. Pearce); Image 54 – Guardian (Zemichael/ Sisay); Image 55 – Zemichael and Sisay (AIT Stock Photo,

copyrighted); Image 56 – Jim and Sherri expedition (AIT Stock Image, copyrighted); Image 57 – Dr. Jim and Sherri Rankin (T. Moore).

In Appreciation: Misgana Genanew, Gondar Hills Resort, Gondar, Ethiopia; Getachew Tekeba, Mountain View Hotel, Lalibela, Ethiopia

CPSIA information can be obtained
at www.ICGtesting.com
Printed in the USA
FFHW022109200819
54430476-60115FF